The Complete Guide to

Organic
Livestock Farming

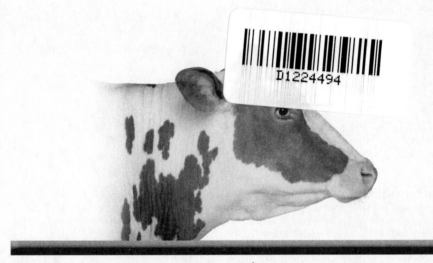

Everything You Need to Know
about Natural Farming
on a Small Scale

Terri Paajanen

THE COMPLETE GUIDE TO ORGANIC LIVESTOCK FARMING: EVERYTHING YOU NEED TO KNOW ABOUT NATURAL FARMING ON A SMALL SCALE

Copyright © 2011 Atlantic Publishing Group, Inc.
1405 SW 6th Avenue • Ocala, Florida 34471 • Phone 800-814-1132 • Fax 352-622-1875
Website: www.atlantic-pub.com • E-mail: sales@atlantic-pub.com
SAN Number: 268-1250

Library of Congress Cataloging-in-Publication Data

Paajanen, Terri, 1971-
  The complete guide to organic livestock farming : everything you need to know about natural farming on a small scale / by Terri Paajanen.
     p. cm.
  Includes bibliographical references and index.
  ISBN-13: 978-1-60138-381-5 (alk. paper)
  ISBN-10: 1-60138-381-9 (alk. paper)
  1. Livestock. 2. Animal culture. 3. Farms, Small--Management. I. Title.
  SF61.P33 2011
  636--dc23
                              2011022358

Printed in the United States

PROJECT MANAGER: Gretchen M. Pressley • gpressley@atlantic-pub.com
PROOFREADER: C&P Marse • bluemoon6749@bellsouth.net
INTERIOR LAYOUT: Antoinette D'Amore • addesign@videotron.ca
COVER DESIGN: Meg Buchner • meg@megbuchner.com
BACK COVER DESIGN: Jackie Miller • millerjackiej@gmail.com

Printed on Recycled Paper

A few years back we lost our beloved pet dog Bear, who was not only our best and dearest friend but also the "Vice President of Sunshine" here at Atlantic Publishing. He did not receive a salary but worked tirelessly 24 hours a day to please his parents.

Bear was a rescue dog who turned around and showered myself, my wife, Sherri, his grandparents Jean, Bob, and Nancy, and every person and animal he met (well, maybe not rabbits) with friendship and love. He made a lot of people smile every day.

We wanted you to know a portion of the profits of this book will be donated in Bear's memory to local animal shelters, parks, conservation organizations, and other individuals and nonprofit organizations in need of assistance.

– Douglas & Sherri Brown

PS: We have since adopted two more rescue dogs: first Scout, and the following year, Ginger. They were both mixed golden retrievers who needed a home.

Want to help animals and the world? Here are a dozen easy suggestions you and your family can implement today:

- *Adopt and rescue a pet from a local shelter.*
- *Support local and no-kill animal shelters.*
- *Plant a tree to honor someone you love.*
- *Be a developer — put up some birdhouses.*
- *Buy live, potted Christmas trees and replant them.*
- *Make sure you spend time with your animals each day.*
- *Save natural resources by recycling and buying recycled products.*
- *Drink tap water, or filter your own water at home.*
- *Whenever possible, limit your use of or do not use pesticides.*
- *If you eat seafood, make sustainable choices.*
- *Support your local farmers market.*
- *Get outside. Visit a park, volunteer, walk your dog, or ride your bike.*

Five years ago, Atlantic Publishing signed the Green Press Initiative. These guidelines promote environmentally friendly practices, such as using recycled stock and vegetable-based inks, avoiding waste, choosing energy-efficient resources, and promoting a no-pulping policy. We now use 100-percent recycled stock on all our books. The results: in one year, switching to post-consumer recycled stock saved 24 mature trees, 5,000 gallons of water, the equivalent of the total energy used for one home in a year, and the equivalent of the greenhouse gases from one car driven for a year.

Dedication

I would like to thank Michael Wilson for being such a wonderful help around our farm while I spent the time writing this book and also my daughter Emily for always reminding me how miraculous Nature is.

I also want to thank the farmers who took the time to take part in the creation of this book by sharing their own stories.

Table of Contents

Chapter 2: Basic Business Considerations...51

Chapter 3: Growing Crops and Forage.......71

Chapter 6: Chickens 121

Chapter 7: Horses 145

The Movement Back to More Natural Products

I n our fast-paced and highly technological world, we are seeing a revival of older ideas that come from a slower and simpler time. People are looking for a more wholesome lifestyle and trying to leave behind the overprocessed foods that have become commonplace on the grocery store shelves.

Over the past two generations, we have traded quality for convenience when it comes to our food. Now, we are coming full circle, and people are looking for quality again. Sometimes, this shows itself in a significant change of diet. Today's natural dietary trends encompass a number of new — or old — eating theories. Some of the more popular examples include:

- **Vegetarian:** Not eating meat (sometimes excluding eggs and dairy as well)

- **Vegan:** Not eating or using any animal products whatsoever (all forms of meat and dairy)

- **Localvore:** Choosing food products that are grown locally rather than shipped from other areas

- **Raw food:** Leaving foods as close to their natural states as possible so as not to diminish their nutritional value (though some cooking is allowed up to a varied maximum temperature)

- **Paleo eating:** A more complicated diet plan that relates to how humans would have eaten before the advent of processed food, which includes grains and baked goods

- **Slow food:** Avoiding convenience food in favor of cooking foods from scratch

- **Organic:** Avoiding any foods grown with chemical pesticides or antibiotics

Other trends involve avoiding certain types of food with ingredients seen as potentially unhealthy, such as sodium, fat, processed sugars, or wheat gluten.

An organic diet is fundamentally different from the other dietary options. There are no restrictions on the specific foods you can eat on an organic diet, but instead, you are making choices to eat foods that do not contain chemical additives. An organic diet is often the easiest to implement because it does not force you to change your overall eating habits. Most organic consumers find their only limitation is what types of organic foods are available for purchase in their areas. An organic lifestyle is sometimes blended with other diet concepts, and sometimes, it is a diet choice of its own.

The momentum of the organic movement is quickly gaining steam, and those who maintain this lifestyle are no longer considered unusual. According to the Organic Trade Association (**www. ota.org**), sales of organic food products went from $1 billion in 1990 to $24.8 billion in 2009. As more farmers and food manufacturers choose to go organic, there are an increasing number of organic products available for the consumer to choose from.

Why is This Trend Occurring?

There is no single reason why people are moving toward organic foods and produce. Different people make their choices for different reasons — or, sometimes, for a combination of reasons. The biggest motivator for choosing organic foods is the health concerns associated with traditional or commercially raised food. This applies to food crops, as well as livestock raised for food. Public awareness about the chemicals our food is exposed to has finally opened concerned consumers' eyes.

When major newspapers are running headlines about nationwide food recalls due to contamination, people begin to question the safety of their food. Consumers have all become familiar with E. coli and salmonella contamination, and even though these particular issues are not directly related to the organic/traditional food debate, they have forced consumers to sit up and take notice of what they put in their bodies. More questions are being asked, and many are unhappy with the answers.

There are also ethical reasons for choosing organic products. Most livestock are treated poorly in conventional agricultural practice; they usually are forced to live indoors in crowded, dirty conditions. Organic farmers tend to treat their animals better and provide better living conditions because they are unable to use antibiotics to counter disease. It is also part of the organic standard to provide humane care and access to the outdoors. Consumers who want to encourage better conditions for livestock animals are looking to organic producers.

With the rampant use of antibiotics to prevent animals from getting sick, many livestock diseases have mutated in order to adapt. This leads to heavier dosing so the antibiotics will continue to work or the introduction of new medications to take the place of the newly ineffective ones. Some animal diseases can also infect humans, so any new aggressive strains can impact consumers' health — not just livestock health.

Some concerns about conventional farming practices go beyond the actual animals' and consumers' health. Concerns for the health of the environment as a whole is also part of the new trend. Constant dosing of the ground with pesticides is beginning to contaminate groundwater and disrupt local ecosystems. Excessively using synthetic fertilizers creates toxic runoff that is fouling rivers and water systems. Moving from a chemical-based farming system to an organic one will benefit the planet at large, as well as the individual consumers and animals.

What does it mean for the consumer and the farmer?

This developing trend toward organic food means consumers are expecting more choices in their food products and farmers are beginning to take notice of that expectation. At first, the idea was new and the organic industry was fragmented. There are now enough consumers and producers who are interested in purchasing organic food that a viable organic market has emerged, and that market is growing quickly.

For the consumer, choosing organic can mean grocery shopping takes a little more time as you search for brands that offer the organic choices you are looking for. In addition, when you do find the products you want, you will probably have to pay more for them.

Finding products with the official USDA organic label will not mean that the product is 100 percent perfect. There can still be potential risks and other problems with organic food. For example, produce or meat can still be subject to fecal contamination from unclean conditions even if the products come from an organic farm. You can be certain that your food is not filled with artificial chemicals, which is definitely a step in the right direction.

For the farmer, managing an organic farm does have unique challenges. You will have to abandon many conventional farming techniques and conform to the strict government guidelines that govern organic produce and meat. Inspections and recordkeeping will become a big part of your farm's life. Even so, it can provide small-scale farmers with a niche market that can set them apart from their nonorganic competition.

Organic versus Traditional Farming

The term "organic" has been bandied about so often in the media that it has started to lose some of its meaning. Although there are some specific parameters to the word when it comes to legal issues and labeling, to the average consumer, it means "produced without any synthetic chemicals or additives." This would also include radiation treatments and any genetic modification, which are generally not part of a small-scale farm operation anyway.

Chemical exposure can come from a number of sources, including fertilizers and pesticides. You might think these only apply to planted crops, but any hay or grain that has been treated with chemicals will pass that contamination on when fed to livestock. Therefore, for livestock to be considered organic, they must eat only organic food (either certified commercial feed or graze in a certified organic pasture). Specifically for animals, the next chemical culprits are artificial hormones and antibiotics.

Those parameters cover the strict definition of organic, but there are other differences between organic and traditional farming practices. As a consumer, you have likely noticed that the largest difference ends up being the cost — at least this is the difference that you can identify tangibly.

Without the added protection of the pesticides, fertilizers, and antibiotics, organic farmers have to deal with more losses compared to conventional farmers. It also can mean more hours must be spent to keep their crops and animals healthy. These added costs end up being passed on to the consumer, which means higher prices at the checkout. Many people who want to change to organic foods are willing to pay the added price, so organic farmers have been able to maintain their market while still managing their costs.

There is a general misconception that organic simply means a label in reference only to chemical exposure. However, part of the regulation also states that animals must have access to the outdoors. It is not a major part of the organic definition, and "access to the outdoors" is fairly vague and undefined as a guideline. Nonetheless, that can be an improvement to the confinement practices of traditional farming.

Aside from technical definitions and specific farming techniques, the biggest difference between organic and traditional farming is the mindset and attitude of the farmers and the customer. Many traditional farmers use practices that will help them run an efficient farm and maximize their profits. Unfortunately, this does not always mean better product for the consumer or better treatment for the animals. On the other hand, organic farmers are usually more concerned with quality. That is not to say organic farmers are indifferent to financial aspects of their farms. They are just as interested as anyone in running a successful business that earns a profit. However, their overall attitude is different, which can be appealing to customers.

In order to protect consumers, only those produce sellers who have obtained USDA organic certification are allowed legally to label their products as being organic. You cannot avoid the issue by making the claim that your products are noncertified organic. The proper labeling is required to make the distinction between organic and traditionally grown foods truthfully. *For more specific clarification on what "organic" means in terms of legal USDA requirements, see Chapter 2 on the legalities of organic farming.*

Overview of this Book

This book will take you to every corner of an organic farming business while keeping it all sized down to a small-scale operation.

The first four chapters will introduce you to the general concepts of small-scale farming and cover the basic topics, such as housing, fencing, feeding, and animal breeding. In this section, you will find some business information you will need when starting your farm. Understanding the legal requirements of an organic farm, as well as all the tax and financing concerns, is vital if you want a business to succeed.

After the general material, several chapters are dedicated to each species of common livestock animal. Each one has its own unique requirements; so do not stop reading with just the overviews. Whether you plan to raise sheep, chickens, goats, cows, horses, or pigs, all of the information you need is here. There is also a section on some less common animals you may want to consider for your farm. Throughout the book, there are stories from real farmers who share their personal experiences running an organic farming operation. These studies can give you a helpful look into what to expect with your own farm.

Because the theme of this book is not simply farming, but rather organic farming, there is a heavier focus on the aspects of keeping livestock that involve organic practices. This means feed and medication will frequently be mentioned.

If you are interested in joining the wave of the organic industry, this book can be your first step down that path.

Chapter 1

Livestock Farming Basics

Any animals kept for useful domestic purposes, excluding those kept as pets, are classed as livestock. The term is usually applied to animals kept in a farm environment. Livestock can include cattle, goats, sheep, horses, pigs, or poultry. The exclusion of pets means that cats and dogs are not considered livestock, even though they might have their own uses around the farm.

The definition does stipulate that animals need to have a use in order to be considered livestock, but some people argue that you

can use the term for traditional farm animals you are keeping as pets. In other words, if you keep a few goats, but only because they are fun to have around and you do not intend to milk them or slaughter them, you still would be able to call them livestock.

Nontraditional farm animals — alpacas, emus, buffalo, bison, or elk — also are considered livestock. As long as they are serving a useful domestic purpose, they are livestock. Rabbits tend to be considered pet animals, even when you are keeping them for meat or fiber. You might want to check with your local or state regulations on this distinction. Some districts might classify hives of honeybees as livestock as well.

Knowing how to define livestock is not a trivial matter. When applying for loans, grants, zoning changes, or permits, you need to be able to accurately describe your farm and its population. If you consider your outdoor rabbits to be pets but the city considers them livestock, you could find yourself at odds with the law if you do not have the proper zoning or permits.

Where you look for this information will depend on the context. To find out about property zoning, your local municipal office should be able to clarify which animals are classed as livestock for your city. For state-level issues, you will want to check with your state's Department of Agriculture.

What the Small-Scale Livestock Farmer can Expect

Anyone wanting to embark on the path of organic farming should know it is not easy or simple. At first glance, all you have to do is raise your animals without any additional chemicals. However, this statement does not even scratch the surface.

If you want to go the extra step of becoming a certified organic farm, you will have to work your way through all the red tape involved and keep immaculate records on every aspect of your farm. Using no chemicals also means you will have to get creative at times to solve problems that only seem to have conventional, nonorganic solutions.

There are many other things to expect as a small-scale farmer, whether you are going organic or not. Even with the smallest of hobby farms, you will have a steady level of responsibility for your animals. You cannot take this responsibility casually. Each day you will have a set of chores to complete, no matter the weather or the time of day. You cannot call in sick when you are ill and expect your animals to understand. They must be fed and given water, and dairy animals must be milked.

However, at the end of it all, you can also expect a sense of personal fulfillment because not only have you raised healthy food, but also you have done so naturally and in a way that is good for the environment. And even better, you can earn a profit while doing so.

Am I cut out for running a farm?

No matter how many books you read, this is a question only you can answer. When considering livestock, be honest with yourself about how you will handle the change in your life. Overlooking any of your own personal issues now will create problems for you later, so be realistic when assessing your abilities and lifestyle.

First, are you in physical shape? You do not have to be a bodybuilder, but physical strength is an important part of farming. Large bags of feed are heavy, as are the tools and equipment you will lug around your property. You might also need to restrain or lift your animals at some point.

Keeping livestock will require a significant time investment, almost certainly more than you expect. Half an hour in the evenings to feed your animals is not going to be enough.

Do you like having a clean yard or well-manicured property? If so, you are definitely not going to be happy with animals. You will need patience when dealing with the mud, manure, vermin, dust, dirt, and bugs.

Are you able to think on your feet to handle the unexpected? Not everyone likes a project that is full of surprises, so think about whether you can deal with constant interruptions, problems, and occasional disasters. If you love a challenge and the idea of handling something new all the time, then this might be the path for you.

You also need to remember the final step of keeping livestock: the eventual slaughter. You will have to distance yourself emotionally from the animals you keep enough to have them killed without being overcome with grief. This is particularly important if you have small children who will be involved with the animals. Thankfully, if this part scares you, you can always opt to keep livestock with the intention of harvesting other products without killing for meat. Goats or cows can be kept for their milk, sheep can be used for their wool, and chickens will give plenty of eggs without any deaths in the process. Another alternative is to raise your animals for live sales to other farmers who are looking to buy livestock.

Farm Planning

Now comes the fun part of starting your farming operation: planning how you are going to use your space for animals, crops, and buildings. It is not as simple as marking off a series of areas for each use at random. Expect to put some real thought into this because it is not easy to move a barn if it turns out you built it in the wrong place. Not only do you need to work out the most efficient layout, but you also need to decide how much space you can afford to give for each function.

Be prepared to make some compromises. No property is perfect, and it might turn out that the best place for the barn is also the best place for the vegetable garden. Shuffle things around to get the best overall layout without worrying about any one area being perfect.

Placement of crops

You do not necessarily have to grow crops because they might not suit your needs or purposes for your farm. If that is the case, you can jump ahead and start planning the placement of the buildings.

Most farmers who grow their own animal feed will usually grow vegetables, hay, or grains, depending on the needs of their animals. *Chapter 3 elaborates more on which crops are suitable and how to grow them.*

Whether you are growing vegetables for your own use or for grains to help feed your animals, all crops are going to need a sunny location. This should be your first concern when planning a garden. Although a few crops tolerate shade, the majority of garden plants need a good amount of sun every day to produce a good harvest. It might seem convenient to tuck your garden behind the house or barn, but that will only work if the sun comes around the building for most of the day. Also, look out for large trees. If you are planning your garden's location in the fall, do not forget that nearby trees will put off more shade when they are in full leaf.

After you have found the sunniest spots, your next consideration should be access to water. Depending on your climate, you might be watering your plants several times a week during a hot/dry summer. Having water on hand will make these chores go a lot smoother. Is your proposed garden location within reach of a hose? If not, be prepared to haul water by hand. A few well-placed water barrels can make a big difference in that case. In addition, do not forget the problems you can have with too much water. A low-lying area might flood during the spring or in a heavy summer rainstorm, washing out your garden space.

Finally, decide how much space you want to allocate to your crops. Decide what you want to plant, how much you want to grow, and how much space your crops will need. Do not get overwhelmed with all the variations; you do not have to calculate it down to the inch. Make a rough estimate, and plan around that.

A good location will get sun throughout the day, have close access to water, be relatively level, and grow near enough to your home that you do not forget about it and let the weeds take over. You are more likely to spend a few minutes tending the garden if you can see it. Locating your garden near the house or where people frequently walk will also help deter some larger pests, such as deer or rabbits.

Placement of barns and other facilities

Your plants depend on the sun, so you will need to figure out their placement first because the sun's location is not negotiable. Next, you can work on where your animals will be housed.

There should be reasonable vehicle access to your main barn, ideally, so a vehicle can drive right up to or inside the building. When moving animals or bringing in supplies, you will want to have direct access without having to park too far away.

Connections to electrical service are important, though not completely critical. If there is no way to place the barn where electrical service can be provided, you might be able to make do. However, if possible, make sure your barn or other buildings will be built where it is easy to connect to an electrical supply. Running electrical wires over several acres of land will become extremely expensive quickly. If the location is too far for conventional electrical service, consider a few solar panels. You will not need as much power for a barn as you would for a house, and it can actually be cheaper than trying to run service a long distance. The American Solar Energy Society can get you started if you are thinking about going the solar route. This organization's website (**www.ases.org**) explains the basics of solar power and has a directory for locating a company to install a system for you.

When considering your need for electricity in the barn, know that you will want power for lights and any electrical tools or equipment you might need in and around the building. It will also be necessary if you are considering electric fencing for your animals.

Your pasture areas need to be adjacent to the barn so that you can move animals in and out without having to escort them from the building to the fenced-in areas. Plan your barn's location with that in mind. The actual design and layout of your barn will be covered further in the next section of this chapter.

The fenced-in pastures and fields will need to be arranged as well. Having one big field is fine, but it might not be the most efficient use of your space. The specific space and pasture needs for each type of animal will be covered in their own chapters, but in general, you will want to have a few divisions in your pasture to keep things organized. Being able to keep animals out of one part of your field will give you the opportunity to reseed it and redevelop its growth potential. You might also need separate field areas into paddocks to keep certain animals apart or to force them closer together in the winter. A few extra fence lines will give you many options.

In addition to the barn, you will want to consider other building needs. A garage for a truck, a tractor, or any other large pieces of equipment that do not belong in the barn is one possibility. Extra sheds for storage should be looked at as part of the overall plan, too. Although you can usually keep most livestock animals together in one barn, poultry, such as ducks or chickens, will need their own housing. Because they can lay eggs frequently, you do not want them wandering around a large barn.

Larger herds of more than one type of animal might warrant separate buildings if you have the space and the desire to do so. Smaller farms are usually more efficient if they have one large barn with separate stalls or compartments instead.

Crop rotation

If you are not familiar with growing crops, you will need to consider crop rotation when planning the layout of your farm. For your garden area, never expect it to be perfectly static from year to year. Parasites, bacteria, fungus, and insects tend to congregate near their favorite food sources. If you grow the same plants in the same plot repeatedly, you will find yourself fighting an uphill battle against the countless pests in the garden.

To avoid this, ideally, plan to move things around each year. If that is too inconvenient, plan on some type of rotation every other year instead. This goes for both garden vegetables as well as any other crops you might be growing for animal feed.

Unfortunately, proper crop rotation is more than just shifting your crops around the garden patch. In order to benefit most from doing this, you should not rotate plants from the same family to the same place the next year. In other words, the insects that love to eat cabbage are just as happy to munch on broccoli. This means that planting broccoli in the spot where you had cabbage last year is not going to do you any good. In this example, you might want to try tomatoes or potatoes in that spot instead. Most vegetables can be replanted back in the same location after three years, so you can put together a three-year rotation plan rather than find a completely new configuration every single year.

3-Year Crop Rotation Example		
Root & Bulb (A)	**Fruits & Seed (B)**	**Leaf & Stem (C)**
Beetroot	Beans	Broccoli
Carrots	Capsicum	Brussels sprouts
Garlic	Cucumber	Cabbage
Kohlrabi	Eggplant	Cauliflower
Leek	Gourd	Celery
Onions	Melons	Collard greens
Parsnip	Peas	Endive
Potatoes	Peppers	Kale
Radishes	Sweet corn	Lettuce
Turnips	Tomatoes	Spinach

Year 1	Year 2	Year 3
A	C	B
B	A	C
C	B	A

For larger plots of grains, such as corn, oats, or barley, your best option is to allow some parts of your field to go unused during alternate years. Doing so gives the soil a chance to rebuild its nutrients. An even better option is to plant a crop of clover, left to be plowed under the next year. Clover is a fast-growing legume that naturally adds nitrogen to your soil, particularly if you leave its roots in the soil when you plow for the next planting. You can either leave some space in your garden empty each year, so you have room to maneuver, or just plan your plant choices so you can shuffle them around. Because you are growing with organic principles in mind, you want to fight insects and pests without resorting to chemical additives. Rotation can be a major part of your overall gardening approach.

Housing Your Animals

The housing requirements for your animals will vary considerably depending on what animals you are raising and the climate you are living in. Livestock will need much different living conditions in the snowy north than they do in the humid south.

In regions where cold winters are the norm, your barns should be as draft-free as possible. The body heat from your animals is usually sufficient to keep the building warm, but a drafty barn is a sure step toward sick animals. A window that faces to the south can help let in sunshine during the winter as well, as long as it is secure from pests or predators.

Where summer heat is a problem, you will have to design your barn with other factors in mind. Ventilation is important, as is ensuring that the barn creates shade during the day. If the barn doors  are facing the south, the sun will stream in during the summer, and even a large barn will not give your livestock relief. Animals that do not have free access to the barn during the day will need some type of sun shelter or lean-to for shade out in the pasture.

As an organic farmer, you should provide more space than conventional wisdom might suggest. This keeps your animals from being crowded, which in turn provides a healthier environment overall. One of the reasons traditional farms must use so many antibiotics is because animals usually live in unhealthy conditions. *The amount of space you need for your animals is outlined in more detail in each of the animal chapters later in the book.* However, in general, you can either lay out your barn so the animals all live in one large pen, or you can divide your barn space into smaller sections. If you intend on breeding your animals, you will definitely want one or two smaller pens where you can isolate newborns and their mothers.

When using a barn is made up of multiple pens or stalls, it is typical to have all the pen doors opening outward into a central walkway that goes through the middle. If the door is as wide as the hall, it will automatically block access and make moving animals around much easier because they will be unable to get past the door even if they manage to get past you.

Your animal buildings will need to protect your livestock not only from the weather but also from predators. Doors and windows must be secure. You must consider whether your animals can get out and whether other animals can get in. A simple lift-latch will not keep a raccoon out of your feed room, and an open window that is not accessible to your livestock can be a perfect way for a fox to gain access to your barn. All windows and vents must be covered by heavy-duty hardware cloth or a metal screen. Chicken wire is fine for keeping larger pests out, but it will not keep out mice.

Barns will need more space than the obvious living areas for the animals. Storerooms for equipment, supplies, and feed will also need to be part of the plan. A separate location for milking animals with a secure spot to harness your animal while you are milking will be necessary if that is your intention.

Winter feeding will require a hefty supply of hay, and those bales are bulky. A hayloft in the barn is a good place to store it, or you could have a separate building for your hay supply. Do not wait until winter to determine where you will store your hay. You will need hay during the other months as well, but it becomes more important in the winter when there is no pasture to be had.

Access to your barn should be easy for everyone involved, both people and animals. Ensure there is a large door so your livestock can access the barn as well as a handy "people-sized" door so you can get into the barn without having to go through the livestock pens.

Space and security are not the only considerations when arranging housing for your animals. You also need to think about the lighting. Having electrical service to your buildings is important. Not only do you want light in your barn for yourself when you are doing chores after dark or before sunup, your animals will benefit from lighting, too. This will be more helpful in northern areas when the days are considerably shorter during the winter. Extending the day length with some artificial light will help prevent your animals from developing any unwanted behaviors associated with boredom or stress.

Housing chickens is different from housing your other livestock animals. They are not kept in the barn but rather their own small house with room for laying eggs. *See Chapter 6 on chicken care for all the details on chicken-specific housing issues.*

Basic Tools and Equipment

Your animals are only half of the necessities of a successful farm. You also must have the right tools. By planning for your needs ahead of time, you will not find yourself missing a simple tool in an emergency. Many farm tools are similar to those you already have around the house for general home repairs, so there is a good chance you will not have to buy many new tools once you get started.

The size of your farm and the livestock you keep will make a difference in how much equipment you will need. Some of these items can be improvised if you are handy and creative, which can save you some money as long as you do not put yourself or your animals at risk with a handmade gadget.

You will need tools to assist in the care of your animals. Plan on purchasing or creating feeders and waterers suitable for your specific type of livestock.  They can range from simple open buckets or troughs to more sophisticated equipment that refills automatically or stays warm in the winter. Being able to safely move your animals around your property is easier if you have harnesses, light chains, or ropes. Shovels and a pitchfork are vital for moving manure and cleaning the barn. Supplies for basic first aid are important as long as you do some research on how to use them correctly.

Depending on how you intend to use your livestock, you also will need to have the right equipment for milking, shearing, and/or butchering. Shearing and butchering are two activities that are only done once or twice a year, and many small farmers will hire someone to do it for them. In that case, you will need no special supplies. Milking, on the other hand, is a daily chore and not something that can be outsourced. You will need little for the actual milking, but storing and processing milk requires more preparation. *See Chapter 12 for more information on operating a dairy.*

Vehicles can be extremely handy around the farm, but they are also quite an expense for a small-scale farm. A pickup truck is the first vehicle most people think of, and it can make a lot of your chores much easier. The obvious choice for moving large animals and bringing home large orders of feed or supplies is a truck. Add a livestock trailer to your truck, and you will be ready for anything. However, a farm with only a few acres and a handful of animals probably will not need this. Your vehicle needs will depend on what type of animals you are keeping. Moving large animals, such as cows, will take nothing less than a truck with a trailer, but if you are only moving chickens, a trailer will not be necessary.

Another vehicle that can be handy on a farm is an all-terrain vehicle (ATV). With a small trailer, they are fantastic time-savers for moving things around on your property. However, ATV will not help you haul things to and from your farm because they are not allowed on roads.

A tractor should be mentioned along with the other vehicles. True tractors (not the "lawn tractors" that are just riding lawn mowers) are extremely expensive and might not be that much help on a small livestock farm. They are definitely helpful when you need to deal with large volumes of manure and for working large plots for crop planting. Though you probably think of tractors when you think of farms, this will not be at the top of your "to buy" list any time soon. That is not to say a lawn tractor would not be helpful. They can be great for small hauling or pulling jobs and keeping the lawn in order.

Without a vehicle to help haul things around, you will have to rely on your own hands and feet. A few good handcarts or wheelbarrows are necessary to move soil, manure, feed, rocks, equipment, and even small animals. A two-wheel wheelbarrow is the most practical. You can carry quite a bit in it, and it is more stable than the older one-wheel type.

Your need for gardening equipment will depend on how extensive your gardening plans are. Basic garden tools include hoes, rakes, shears, trowels, spades, hoses, and buckets. A small rototiller can be helpful to till and loosen the soil for garden areas. Some baskets or buckets for gathering your harvests will be needed too.

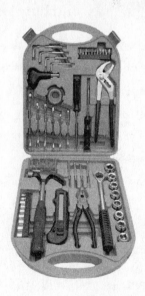

Last are the general repair tools you probably already have around your house. They will become doubly important when you are running a farm. It might be a good idea to have several so you can have handy tools in your home as well as in the barn. A good start for a general toolbox includes a hammer, pliers, an adjustable wrench, a socket set, a drill, screwdrivers, a pry bar, a level, a tape measure, a circular saw, and a collection of nails and screws in different sizes. A chain saw can be a great tool for general cleanup around the property, but it is not essential.

Working with Contractors

There is no reason why your new farming endeavor must be tackled alone. Using outside help is the best way to bring the right skills to your farm and get things done properly without having to master everything all on your own. It also saves you the cost of buying specialized equipment you will only use once or twice a year.

Most projects on a farm can be handed to a contractor, although having someone else do everything will get expensive. Building your barn, running electrical services, putting up fences, or plowing fields are all examples of big jobs you will probably want professionals to help you with. Occasional or annual chores such as cutting down and removing dead trees, transporting animals, baling hay, and butchering can also be contracted out.

Rural communities will have no shortage of workers from whom to choose. In fact, there might be so many people you will not know whom to hire. Ask your friends or neighbors for recommendations first. Hearing good things about a person or company can go a long way in determining who is best for your job. Once you have a few potential people lined up, ask for estimates on the job. Several quotes will help you establish a fair cost before you commit to anything. In addition, do not forget the time-honored tradition of bartering. Someone might be happy to help you put up a fence in exchange for half a cow at butchering time, for example.

Because you are running your farm as a business, you can consider the cost of contractors as a deductible expense when it comes to income taxes. If you are planning on using these types of expenses in this way, you may have to stick with larger contracting companies rather than handy neighbors. Although companies have no problem issuing invoices and receipts for your records, individuals who do these jobs on the side probably will not. If you are paying for the job outright, they will expect to be paid in cash.

Part-time contractors can take a frustratingly laid-back approach to a job, which can try your patience if you are not expecting it. Make sure your schedule is flexible, and just let yourself go with the flow. If you are working on a tight deadline, play it safe and stick with the full-time professionals. You could be in real trouble if you have a delivery of new animals before your fence is finished.

Fencing

Fencing around your fields or pastures is important and will need to be suitable for your animals to keep them safely contained in their proper places. Building a fence that will hold a sheep is not going to be much help if you add in a few big cows or hyperactive goats. Not all fencing is created equal; so do not make assumptions about generic fencing being suitable for anything.

Types of fencing

Each type of typical farm fencing has its own pros and cons, and some can be better suited to certain animals or certain budget limitations. Fencing is not something you want to skimp on — even if it is expensive. A good fence is vital to keeping your animals contained and safe.

At the cheaper end of the spectrum is electric fencing, which can also be the easiest to install. An electric fence is more complicated than standard woven wire, but learning how to install it will not take much time. The wires are charged by a power supply, called a fencer, and provide a painful shock when touched. Because your animals are given this jolt immediately on contact, an electric fence does not need to be built to the same sturdy standards as a wire fence that will be pushed or leaned on. Although the wires will give a shock, the fence usually is not a health hazard or considered dangerous, even with small children living on the farm.

A typical electric fence will have two to four lines of wire at even heights, depending on the size of the animals you are keeping. Large and rather slow cows can usually be contained with just two live wires, whereas a frisky little goat or lamb will need at least four (if not more) to keep them from slipping under the low wire or hopping between the higher ones. When using electric fencing, you will have to clear the fence lines of high weeds or they can make contact with the wires and short out the system. High-wattage fencing units can handle some contact with weeds or branches because they can compensate for the power that drains away from the contact with the ground. They will not short out, but if your wires are in contact with any obstruction for

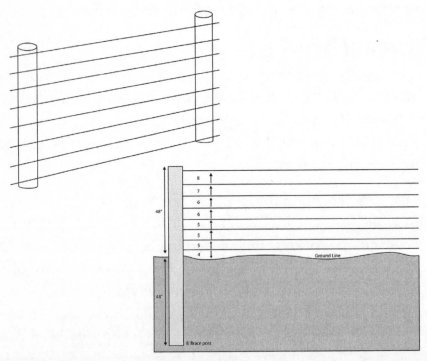

Example of an electric fence designed for goats

too long, the extra power usage will get costly. Because of this, it is best to keep the fence lines clear.

For a more no-nonsense fence, you can always go with the traditional woven, wire-style fence. They are durable, long lasting, and as simplistic as something can get. Chain-link is the strongest, but it is usually much too expensive for farming purposes. For closing in pastures, farm-grade woven wire works best. It comes in long rolls in widths of 4 to 6 feet. Just fasten it to your fence posts, and you are done. Unlike the electric fencing, the posts must be heavy and sunk deeply to withstand the pressure of your animals testing the fence. A cow can put a lot of weight on a section of fencing, and the posts must hold up.

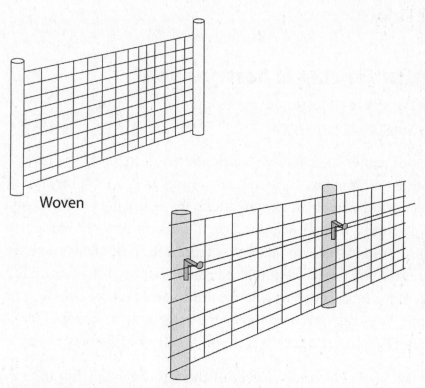

Woven

Examples of woven wire fences. Bottom illustration showing a woven fence with an electric wire at shoulder level.

Rail fencing is fine for show, but it will not contain any animals, though putting up a rail fence along with some electric wires makes a good combination. The rails make the electric fence line more visible.

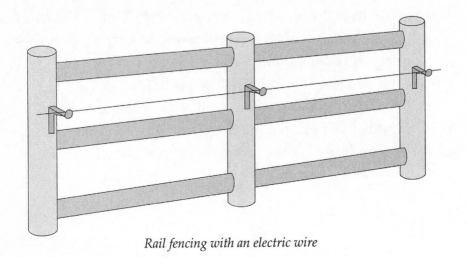

Rail fencing with an electric wire

Which fencing is best for you?

There are several factors to consider when deciding which type of fencing your farm needs.

When it comes to cost, the electric system is actually the cheapest. Even though you have to buy a fencer to power it, the cost of the wire and wire insulators to attach the wire safely to the posts, will come to a much lower total than the same amount of regular fencing. You do have to take into account the cost of electricity to power it, which will continue to be an expense for the entire life span of the fence. For small fields, this literally can be pennies per month. You can eliminate this completely with a solar-powered system. For the farmer on a tight budget, this is the way to go.

Cost savings alone puts electric at the top of the list, but not everyone wants to deal with the complexity of running the wires and making sure the current is flowing evenly over the entire fence line. Also, larger pasture areas might be outside the reasonable capacity for an electric fence. A field that is too far from electrical service will also need a conventional fence.

One final thing to consider with electric fencing is that the wires do not actually create a physical barrier. If there is a power outage or something falls on the lines to break the current, your animals are not going to be held in place. Without the shock, the fence will not work. In areas where storms and power outages are common, this type of fencing might be more trouble than it is worth.

That said, animals you have had for a few seasons will not realize the current has disappeared because they already know to stay away from the fence lines. They are not going to test the wires and will not even know the shock is gone.

No fencing is perfect, and a determined animal can go through just about anything if it wants to. This is particularly true of large livestock, such as cows or horses. The best way to have effective fencing is to make sure your animals are content on their side of the fence so they are less likely to wander. Hungry, bored, or crowded animals are more likely to want out. Happy animals are more content to stay put.

Butchering and Processing Basics

Butchering can be a difficult subject, particularly for new farmers who have not quite gotten used to the idea of having to dispatch their own animals. Some people have no trouble, and some never become comfortable with it.

Luckily, you do not necessarily have to tackle this on your own. In fact, most small-scale farmers choose to use a slaughterhouse service for butchering rather than do it in-house. Being able to legally sell meat to the public requires that you have the facilities to do so cleanly, which means there are government regulations and inspections involved. For the small farm, it is not usually worth the money, time, and effort to maintain all of that.

It also can take a lot of time to kill and process a number of animals, even on a small farm. Letting a professional take over the chore can streamline your own operations, particularly during the fall season when you are already busy getting everything ready for winter.

A slaughterhouse usually charges by the pound of meat or sometimes charges a flat fee for a certain animals. The cost is low enough that it should not adversely impact your final profit margin as long as you account for it when determining pricing. Some operations might even offer to pick up your animals, but it is more common for the farmer to deliver them to the facility. In order to minimize the stress on your livestock, plan your appointment ahead of time so they are expecting you. This allows the workers to handle your animals immediately and saves the animals the anxiety of being kept in a crowded holding pen while they wait.

Overview of procedures

For those who have decided to do at least some of their own butchering, the following is a general outline of what is involved. *Specific details about meat cuts for each animal can be found in their respective chapters, but no further butchering techniques will be provided.*

The first step is slaughtering your animal. Isolate each animal from the herd before doing this. It will upset the others to see one of their companions die unless you are slaughtering pigs. Pigs have no reaction at all and will ignore the activity. A common kill method is a gunshot to the head. A .22 caliber rifle is sufficient, and the shot needs to be either between the eyes or right at the temple. It is quick, painless, and clean. If you do not have weapons around, a bolt-gun designed for animal slaughter is your next best choice. It will not kill the animal but will knock it instantly unconscious so it can bleed out.

After you have killed or stunned your animal, you need to let it bleed out. A clean slice across the neck will open up the jugular vein and allow the blood to drain out quickly. You will have to suspend the animal by its feet while this is going on. Large animals can be difficult to manage at this point, so plan on having some helpers around.

Next, you will need to skin your animal and remove the unwanted internal organs. Care needs to be taken at this stage so you do not cut open any intestines or other areas that would foul the meat. Once the carcass is prepared, next comes the butchering. Every animal will need to be cut as per its own anatomy, so make sure you study a few diagrams before getting out the knife. However, if you do not butcher your animal with the precision of a professional, no harm will be done.

Before you start, be prepared to handle and dispose of the waste material (organs or skin). In order to keep your work area clean, you cannot have piles of offal lying around.

Because chemicals are not used in butchering anyway, there is little difference between traditional methods and organic ones. The biggest difference is the attitude of the farmer toward his or her animals. Even at the end of their lives, your livestock should be treated kindly. Be as gentle as possible during the process so they are not fearful. Not only is it ethical, but it also does make the chore easier if you are not trying to handle a skittish animal.

Butchering and processing guidelines

You will have to ensure your facilities meet the legal standards if you are going to sell your own butchered meat. If you are not using an outside slaughterhouse, you must comply with USDA regulations for managing a legal slaughterhouse. *The specific rules and regulations involved in legally selling meat are covered in Chapter 2.*

These regulations are beyond the scope of this book, though they should not be overlooked if you are going this route with your farm operations. The U.S. Department of Agriculture (USDA) website has all the documents you need to outline what you must do and how to go about getting your farm inspected. Specifically, the Food Safety and Inspection Service (FSIS) handles this type of regulation, and their website (**www.fsis.usda.gov**) should be your first stop. In particular, the Federal Grant of Inspection Guide (**www.fsis.usda.gov/regulations_&_policies/ Grant_of_Inspection_Guideline/index.asp**) is extremely helpful and includes both a list of steps and the application form.

Feeding Your Animals

Aside from avoiding certain medications, food is one of the biggest components of managing an organic farm. You will need to pay close attention to this area of your farm operations not only to make sure your animals are fed nutritiously but also to ensure their food conforms to organic standards.

What type of feed will I need?

This will vary greatly depending on your livestock animals. Do not assume all animals will require the same food, even those that seem similar at first glance, such as goats and sheep, for example. In order to keep organically raised animals as healthy as possible, you should give them the best possible feed rather than generic feed that is "good enough." Knowing the right nutritional requirements for each animal is crucial. Check each chapter for the specifics on the nutritional requirements of each animal.

Generally, your animal feed will be made up of some proportion of grass and/or hay and grain. Ideally, you would want your animals to get the most of their food naturally from foraging out in the pasture. That means less feed storage for you and better nutrition for your animals. *You can find more on grass feeding later in this chapter.*

Organic horse feed pellets

However, when it comes to the feed you must store and provide for your animals, you will be using hay and grain. For some animals, you can buy commercially prepared feed in pellet form, but always be careful to buy only organic feed. It can be harder to find and will cost more than conventional feed formulas.

Grains should not make up the bulk of your animals' diet because they are not nutritionally suitable and will add quite a bit of fat to your animals instead of lean meat. Grains make an excellent treat and can be offered when you need to lure your herd back into the barn at nights.

Buying feed

Bags of commercial feed and grain usually are available year-round from any agricultural store or local farm co-op. Delivery usually is available for large orders, or you can haul your own. This is one time when a pickup truck will come in handy.

Hay might only be available during haying season, which is usually around late summer, so watch local ads to see when farmers are selling. You might be able to buy hay from a central retail location, but typically, you will be buying right from a farmer. Bales of hay are just too large and bulky for any reasonable transport and storage, so stores do not manage it. This can make it a little less reliable than bags of grain, so when you find a good source for dry hay, make sure you stay with it. Farmers sometimes try to sell all their hay immediately so they do not have to store it, which will leave you with few options if you decide you need a few more bales in the middle of winter. Some will keep some bales on hand so they can sell it at a higher price at these times. If you are lucky, you will have a local hay supplier who does not mind delivering throughout the year so you do not have to store it. You will probably have to pay for it up front though.

As with commercial grains, your hay must be organic. Check with the grower to make sure his or her hay is organic and ask to see certification if necessary. Hay growers are exempt from needing certification if they gross less than $5,000 per year. This might be the case if you are buying from a neighbor or someone local. Unfortunately, if you are going to manage a certified farm, you will need to buy hay from someone who also is certified so you can maintain the necessary documents for your own certification.

Storing feed

Keeping your livestock feed safe is almost as important as keeping your actual livestock safe. You will probably have just as many threats to worry about, so do not take the issue lightly. Food that has gotten moldy or is contaminated with mouse droppings is going to put your animals at risk, and any organic farmer who is trying to avoid antibiotics and medication must keep this in mind.

Loose grain or feed pellets should be in tightly closed containers made of any material that cannot be chewed through. It will also need to be watertight as well. Metal or food-grade plastic will work great for holding feed, although even heavy plastic can be chewed if a rat or mouse is industrious enough. If the containers are plastic, inspect them periodically. Concrete is sturdy, but it is not watertight unless lined inside with plastic.

Lids need to be secure to keep out animal pests. If there is a chance raccoons might get into your food storage area, a locking lid will be necessary. They can solve most latches without hesitation. Storing your feed in a secure part of the barn will help.

Hay will be a bit of a challenge on a small farm because it will take up a lot of space, and it cannot be stored outside unless you can completely cover it with waterproof tarps. Even if protected from rain, your supply will still be at risk from mouse infestation. Therefore, you should store it indoors if possible. As mentioned previously, you might be lucky enough to have a source that will deliver a small load at a time so you do not have to find storage for too many huge bales of hay.

Either way, you do need an indoor storage location. A loft in the barn or a separate building might be required. It must be dry but also open enough to allow for ventilation. Even thoroughly dry hay can be prone to mold and mildew, which can lead to a substantial loss of feed or even ill animals if they are fed moldy hay by accident. Wherever you keep your hay, it needs to be well ventilated as well as protected from moisture.

Grass and forage feeding

Grass feeding is the newest trend in organic livestock, and means animals have a fresh and healthy diet just as they would in nature. All you need to do is let your livestock roam around their paddock and eat what is growing, though the size of your pasture space will determine how practical this is going to be.

This approach can cut down feed costs tremendously and will reduce your own chores if you do not have to go out to the barn every day to unload hay. It is also a more natural approach that appeals to many farmers. Grasses and weeds that are still growing are a much more nutrient-rich source of food for your animals than a bundle of dried hay. When the water is removed from plant material, a lot of the vitamins and minerals are lost as well. With a hay-only diet, you have to compensate for this with supplements. Although this is allowed under organic standards, you can only use natural vitamins. No synthetic products can be used.

Unless you have a large area relative to your herd of animals, it is not likely you will be able to rely on grass or forage feeding 100 percent of the time. If you have year-round growing weather, you will have better success. Otherwise, you will be forced to rely on grain and/or hay through the winter and early spring months.

The average field, when left on its own, will grow a wild and varied mix of plants. This might be suitable without any extra work from you for animals such as goats, but cows are more finicky. They will only eat certain things. In that case, you will get the most food per square foot if you seed it for the right plants. Keeping your animals to one part of the field will allow you to seed the unused space so you will have the right forage plants growing. *Chapter 3 has more on managing a field for forage crops.*

Depending on your animals, watch for any toxic plants that can crop up in your pasture. This varies from species to species, so you must do your research before turning your animals out into any pasture, whether you are adding extra feed to their diets or not. Unless you are using small paddocks that have been trodden down to dirt, always check for dangerous plants. Some of the more obvious ones are foxglove, cockleburs, nightshade, and poison ivy. Even goats, with their famed cast-iron stomachs, need to be protected from certain toxins. Know your animals and what they must refrain from eating. It might seem impossible to safely identify every plant in a pasture, but to keep your animals safe, you do need to make the effort.

Foxglove

In many cases, poisonous plants taste bad enough that your animals will not continue eating past the first nibble.

Nightshade

Poisonous plants soon begin to show themselves because the grass around them is eaten and one plant will stand alone. That is when you should investigate. Do not assume you can leave it growing because your animals have ignored it. Dig it up and dispose of it where it will not be eaten. In order to keep organic standards, no chemical sprays can be used to fight any problem plants.

Initially, the biggest challenge in starting your organic livestock farm will be finding organic sources of feed for your livestock. Once you have established a good source for all your feed, you can make your choices based on what is best for your specific animals throughout the year.

CASE STUDY: SPECIALIZING IN A BIT OF EVERYTHING

Katia Holmes
Misty Brook Farm
P.O. Box 62
Hadwick, MA 01037
www.mistybrook.com

Located near the Quabbin Reservoir in Massachusetts, Misty Brook Farm is a multipurpose operation that offers milk, eggs, and many kinds of meat.

The farm started with a handful of cows, chickens, and one boar and sow. Now the farm is raising hundreds of animals every year, including Jersey cows and Tamworth pigs for milk, beef, and pork. Chickens are kept as layers as well as for meat. You can really find a bit of everything at Misty Brook Farms.

All of the products are sold directly to the consumer. Raw (unpasturized) milk is available directly from the farm shop, and their other products can be found for sale at many nearby farms and markets. Vegetable crops are sold through a community-supported agriculture (CSA) cooperative. *There will be more on CSAs in Chapter 13.*

Holmes did not find the certification process to be difficult, though she notes that there is a great deal of paperwork and record keeping involved. With some networking and research, she has found good sources for all the necessary organic feeds and supplies. Her animals are so naturally healthy that they seldom need to use a veterinarian. When they do call the vet, even he is impressed with the rapid recovery her animals make.

Like most small-scale farmers, she does take their animals off-site to be slaughtered and butchered by a professional. Other than that, they do all the work themselves. Much of the heavy work is even done by draft horse.

Holmes said they chose to become certified because the ideals of organic farming matched her own when it came to how she wanted to care for her animals. Her experience working on a number of conventional farms before founding Misty Brook Farm helped her tremendously when making decisions. She recommends working on farms as an excellent way to gain both knowledge and experience. Staying flexible and diversified is her main suggestion for anyone just starting out.

Chapter 2

Basic Business Considerations

A small-scale farm might seem different from traditional business, but do not overlook any of the business aspects of it. You will be more successful in the end if you treat your farm like a business rather than a hobby.

Most farmers recommend discussing the legal and financial aspects of your business with a lawyer and an accountant before proceeding. The U.S. Small Business Administration (SBA) is a vital source of assistance and information, and it is a good idea to check their website often when you first begin planning your farm. The website (**www.sba.gov**) offers information on planning, legal issues for small business, taxes, and many other aspects of your business. They also are involved in providing financing and can be a source of startup loans.

Planning for the Size of Your Business

The biggest consideration for the size of your farm is the size of the property you will use. If you already have land, your farm will not be able to grow beyond that unless you plan to purchase more. Unlike conventional businesses, the size of your farm is literally measured by the amount of land you have. You will not

be able to raise any more animals than your space will allow. If you are on the market to buy land first, then you will have to decide how much land you can afford.

Spending all your money on a huge spread of land will not be helpful if you have no finances left over to build a barn, so do not dream too big. A piece of property large enough for some expansion without breaking the bank should suit fine.

With some research, you can see how much money you can make with a certain number of animals. This will depend heavily on your local market and potential customer base, so there are no set guidelines. However, you should look into the cost of organic meat or eggs in your area and use that as a rough guide as to what the market value is. If there are no sources nearby, check the prices in the next closest major city. Once you know what produce is selling for, you can establish a rough profit potential for each animal. Of course, you also will need to know how much meat or milk you can expect to get from each animal. *This infor-*

mation will be provided in the animals' respective chapters throughout this book. These chapters also will provide some estimates on how much feed each animal will eat, so you can determine what your costs will be as well.

Decide what type of profit you hope to achieve in exchange for your time and effort. If you go too small, you will not earn enough to make it worthwhile. One benefit of staying small is that you stay "under the radar" for some regulations. If you are grossing less than $5,000 per year, you can get away with not having your farm inspected by the USDA and still label your products as "organic" — though not with the official USDA label or logo. When going this route, you are expected to still follow the official organic standards and can find yourself audited by the USDA if they suspect you are being dishonest with your claims. Staying small is something to consider if you want to farm only as a part-time project. However, to really earn a living, you will need to accept the red tape and plan to be larger.

You will not be able to plan the size of your farm fully until more of your overall business plan has been put together. Once you know more about the profit potential of each animal and likely markets for customers in your area, you can determine more accurately how large you should aim to be.

Legal Entities

As with any business, creating a separate "entity" that is distinct from yourself and your family is important. This can help protect you from losses and liabilities and make it easier for your business to get financing and tax breaks. It need not be complicated. There are four options: sole proprietorship, general partnership, limited liability company (LLC), or a corporation.

A sole proprietorship is the simplest and least complicated of the three and is often the most appropriate for a small farm. You do not need to file separate taxes for your business, and there are no fees to maintain a sole proprietorship. Because this form of business is really just an extension of yourself, you are personally liable for any financial losses the business incurs. There is no protective barrier between yourself and the business. Even so, this is a good place to start for a new farmer until he or she has a better idea of the scale the farm is going to operate on.

A general partnership is created when two or more persons associate to carry on a business for profit. A general partnership is similar to a sole proprietorship in that all parties remain personally liable for any losses. However, those risks are dispersed equally among all partners. You can pool your assets and then share the risks, which makes this a good choice for small-scale operations. If you want to work with others but not take on the personal risks, you will want to move to another entity.

The next choice is the LLC. This falls in between the sole proprietorship and a corporation, and it offers the best of both. The biggest benefit is that your personal assets are not connected to the business, so you are not individually liable for business losses. There is more paperwork involved, and the specific filing requirements and fees will vary by state. This is the best choice for the small farmer who wants to take his or her farm operation beyond the "hobby farming" stage, particularly if the plan is to invest money in equipment, land, and buildings.

The last option is to form a corporation. Generally, this is too big an undertaking for the small-scale farmer, especially at the outset. Granted, there is additional protection for you in terms of liabilities, and there are also more tax breaks for corporations. However, the cost and effort to establish and maintain this kind of business is considerable and often too complicated to handle on your own. If you do want to take this step, find a good financial adviser to walk you through it. You will likely want to keep them on as a regular contractor to keep all your paperwork in order. Ideally, try to work with someone with agricultural experience.

Knowing which form to take can be tricky at the outset. The biggest consideration is whether you want to be personally liable for the business's losses. If you are going to be sinking a large investment into this business or perhaps trying a riskier idea, such as exotic livestock, you will want to protect your personal financial state with at least an LLC. This is usually the deciding factor — more so than the actual profit or earning levels. No choices are set in stone, and most business owners start as sole proprietorships and then work their ways up to corporations only after they have started to take on more financial risks with expansion. Some never go beyond the sole proprietor stage.

Preparing a Business Plan

A business plan is a structured outline of the framework around which you are going to build your business. You will want to summarize what you want to produce, whom you will sell those products to, and how much it will all cost you to operate. The basic goal of a business plan is establish a future path you intend to follow.

A successful business plan will have the following sections to it:

- **Business summary:** Outlines the plan for your business

- **Products and services:** Lists what you plan to offer as a product or service

- **Market analysis:** Compares your intended business with any competitors, outlines what your potential customer base will be, and states where you plan on selling your products

- **Financial planning:** Outlines your startup costs, expected ongoing costs, and projected sales income

- **Business organization:** Discusses the legal form your business will take and any staff or partners who will be taking on certain duties

Trying to plan your entire business before you begin can be daunting. Thankfully, in this age of entrepreneurship, many resources are available to help you. A number of guides with blank templates are also available to help you get your business plan in order. Bplans.com is a helpful website (**www.bplans.com**) that has hundreds of sample business plans, including several in the agricultural industry. The website also has further articles on how to best judge your potential sales, market, and costs.

A business plan is not a document you should put together over the weekend. Serious time and research must go into your business plan if it is to provide any benefit to your project. Books and software can help you work through each individual section, or you can hire a financial adviser to assist you. The biggest benefit of writing a business plan is having a record of all your initial research. After sourcing out organic feed, you will know roughly how much you will spend on food, which then leads to a more informed decision about which animals you should keep. Once you check the local market for produce prices, you can determine your expected sales and whether certain areas of the industry are too saturated in your area. All this information can be overwhelming if it is not organized in a single set of documents.

Your business plan is not just something you create at startup and then put on a shelf. You will use your plan in many situations, including applications for financing help, collaborations with other entities, and obtaining legal authorization from the government. This should be a road map to help you and your business stay focused and on track during those first several years. Each year, amend your projects based on your new current position. If your costs were higher than expected, adjust the next year's balance sheet so you can compensate with pricing.

Finding Financing

If you are starting out small, you might be able to skip this step and launch your farm from your own savings. The areas where you usually will need financing are purchasing land, buildings, and equipment. Animals themselves are relatively inexpensive, and

 you can control your spending by starting with a small herd. Already living on a parcel of land large enough for a farm will make a big impact on how much new financing you will need.

The most conventional way is to get a small business loan from your local bank, although an agricultural bank or co-op might be more sympathetic to your cause. You will need to have a complete business plan and some capital of your own in order to get a loan. Using a bank you already have a good history with can go a long way in your favor as well.

A more generic line of credit is another option and is more closely tied to your personal financial state rather than your intentions with the business. You often can apply for a credit line based just on your credit score and bank history, and this option can provide a reasonable amount of funds with little paperwork. Credit cards also have this advantage, but with their high interest rates, they are not desirable as a main source of startup financing. They do, however, provide a helpful buffer for short-term or emergency funds.

If you own your home, a home equity loan is another source of financing that is not tied to your business potential but rather the appraised value of your home and the state of your current mortgage. Be aware that you are putting your home at risk should the business fail, and you are unable to pay the money back on time.

Private investors are harder to find. This approach does have some benefits in that dealing with one person can be easier than trying to work with a large bank's policies. Someone who believes in organic principles might be swayed more by your attitude than just the numbers on a balance sheet and can be willing to get involved with financing. You will not have the same guarantees as dealing with a bank, and a legal contract should be written up so both parties are covered as things progress. No matter what the attitudes are at the beginning of the project, never assume the deal will continue without any conflicts. A contract that outlines the responsibilities of both parties is necessary so neither one can backtrack or deny what they are supposed to be providing in the deal.

Unlike a bank governed by federal laws, an individual is not a regulated entity. In most cases, if your investor reneges on his or her financial responsibility, your only recourse is to sue for the money.

Large venture capitalists will probably not be a viable option, as they are looking for large profit returns on their investments, and farming is not going to provide that.

Loans and grant programs

Loans and grants are two different things; so do not consider them equivalent sources for money. The state or federal government issues grants with specific stipulations on how the money can be used through their many departments and agencies. When it comes to agriculture, grants, typically, are issued as financial assistance to farms willing to take part in some sort of government program. You might be required to breed a rare species of livestock, use certain ecological conservation techniques, or follow certain protocols. Some states will offer grants to organic farmers simply for being organic, but you will have to check on what is offered in your area. The Grants.gov website (**www.grants.gov**) is a good starting place.

Grants are awarded to suitable farms based on applications, and the money does not need to be repaid. Depending on the rules of the grant, there might be ongoing paperwork to report on your progress or prove you are fulfilling your obligations under the grant. It might seem appealing to get "free" money, but sometimes, the added work involved can cancel out any benefits.

Loans are more conventional and will not need much explanation here. Most banks will have their own loan programs in place for small businesses, and the SBA manages its own set of loan programs offered through local banks. When a bank might not want to extend credit itself, the SBA will be more open to it and can guarantee some loans with your bank if you meet their criteria.

With the SBA, there are a few loan programs from which to choose. The most popular is the 7(a) Loan Program, a flexible loan program designed for new businesses in the startup phase. Within the 7(a) Loan Program, the Small/Rural Lender Advantage (**www.sba.gov/rurallenderadvantage**) might be of particular interest to you as a farmer. This loan program is simpler and more streamlined than the SBA's standard 7(a) loan, which is intended to encourage more small businesses to use the resources in rural areas. A small loan of less than $50,000 will require only a single-page application, and the SBA guarantees nearly all loans less than $150,000.

Cost considerations

If there are places where you can cut down on costs, you should pursue them. Your animals do not have to be purebred, which cost considerably more than crosses or mixed breeds. Unless you are planning on a specific breeding program, your initial herd can be made up of various breeds to save some money.

Equipment can be purchased secondhand, though you always are running the risk of any breakdowns or failures with older items. Non-power items are best for buying used unless you are handy around engines and can fix any problems that crop up with larger machinery. You also can consider renting some items you will not use on a regular basis or even borrowing them if you know your neighbors well enough.

Another cost consideration you need to incorporate into your planning is the fact that you are dealing with living animals rather than inanimate inventory. Death, illness, and many other unexpected issues can come up that will impact your cost structure. Consider these from the beginning by allowing for a few deaths each year, so if it does happen, your budget is already built to accommodate the loss.

Zoning and Permits

Before you can actually begin farming, check whether your property can even be used for this purpose. Only certain types of buildings and businesses can be built on certain parcels of land, and these regulations are called "zoning." Land zoned for industrial use cannot be used to build a house, nor can you start a business on land zoned for residential use. The same applies to farming.

If you are living in a typical residential area, it is almost certain you will not be allowed to create any type of farm on the property beyond a few chickens. Considering how much space you will need for a farm, it is not likely you are going to take that route anyway.

For the sake of expediency, finding true agricultural land that allows for a residence is your best option. A large acreage with a home in a rural area is ideal and not that rare of a find. Of course, there is a difference between allowing a farm for personal use and allowing one that is a business.

The rules and regulations for zoning are city-specific, so check with your local city or town office to find out what you need to know. If you find a suitable property that is not zoned to your needs, you can apply to have that changed. Again, these procedures are unique to each city and cannot be generalized accurately. When trying to purchase a piece of land that is not already zoned properly, factor that into your real estate transaction so the deal is not binding if the zoning changes are not allowed.

Once you have established the proper zoning for your land, you will need to acquire the proper permits from your city so you can keep animals. Again, these regulations are unique to each city, so you will have to check on laws that pertain to your area before beginning. When applying for a permit, you are usually required to provide information regarding the housing and facilities for your animals, the species and number of animals you will be keeping, and the amount of space you will be using for your operation. Permits will have annual fees associated with them, and you might need more than one if you exceed the number of animals per each permit.

Permit requirements can vary by zoning type as well. In other words, you might find that you can get a permit to keep animals in a residential area, but the regulations will be tighter than the same type of permit in an agricultural area. For example, you might be required to keep fewer animals.

Legalities

The biggest challenge to maintaining an organic farm is keeping up with the exact standards and paperwork requirements of the government. You can put your entire operation at risk if you choose to ignore the rules or red tape, and ignorance of the law is no defense.

USDA organic requirements

The government body that regulates the use of the "organic" labeling in food is the U.S. Department of Agriculture (USDA), not the Food and Drug Administration (FDA) as many people believe.

Within the USDA, the National Organic Standards Board (NOSB) and the National Organic Program (NOP) regulate the standards set for organic foods. Without burdening you with further bureaucratic divisions, it suffices to say the official standards for organic labeling are defined by the NOSB.

The official regulations are several pages long and written in legal jargon that can be less than helpful to the average farmer. The following is a layman's summary from a USDA brochure on the NOP:

> "Production and handling standards address organic crop production, wild crop harvesting, organic livestock management, and processing and handling or organic agricultural products. Organic crops are raised without using most conventional pesticides, petroleum-based fertilizers, or sewage sludge-based fertilizers. Animals raised on an organic operation must be fed organic feed and given access to the outdoors. They are given no antibiotics or growth hormones."

In order to have official certification, your farming operation needs to be inspected by the proper certifying agency, and you will need to provide a significant amount of paperwork. According to the USDA website, along with your application for certification, you will have to provide the following:

- The type of operation to be certified

- A history of substances applied to land for the previous three years

- The organic products being grown, raised, or processed

- The organic system plan (OSP) describing the practices and substances used in production, which must also describe the monitoring practices to be performed to verify that the plan is effectively implemented and the recordkeeping system and practices to prevent commingling of organic and nonorganic products and prevent contact of products with prohibited substances

Once certification has been established, you will need to continue to keep detailed records along these lines to satisfy the inspection process. When you apply for certification, your farm will be inspected, and there will be annual inspections after that. There also can be unannounced inspections to verify that there are no prohibited substances being used or no prohibited practices taking place.

Agents for certification are organized at the state level, and you will need to contact your state's Department of Agriculture to arrange your inspections. They should be able to assist you in the application process and answer your questions.

Given the amount of work required to earn that little organic label, it can be tempting to skirt the rules. Be aware that anyone selling a product with a fraudulent USDA organic label can be fined as much as $11,000. It is not worth the risk for your farm. If you do not want to deal with government regulations, you can still practice organic principles without using the term "organic" on your packaging. Other such labeling such as "natural" and "free-range" are not currently regulated.

While you are gathering your documentation for the three years of land history, you can still begin your farming operation as long as you keep it small. As previously mentioned, a producer with gross sales under $5,000 is exempt from certification.

This only covers your certification to becoming an organic farm. There are separate rules and regulations involved in selling any particular product such as milk, meat, or eggs.

Selling milk

Milk falls under the same regulations as any other organic food product, so it must be certified in the same way if you intend to label your milk as organic. Beyond that, rules for milk will vary from state to state. The main consideration when it comes to selling milk is the processing. In many states, it is illegal to sell unpasteurized milk, which is usually just  called raw milk. With the surge of interest in healthy and unprocessed foods, this is becoming a controversial issue. At this time, 28 states allow the sale of raw milk.

Pasteurization is a relatively simple process in which the milk is heated to high temperatures in order to kill any bacteria. Because there is no addition of chemicals or other substances to the milk, pasteurization does not interfere with organic standards. This required step can be a burden to the independent farmer, but small-scale pasteurizers are available if you wish to invest in the equipment for your dairy operation. *The necessary equipment will be discussed further in Chapter 12.*

Some states' laws do make exceptions for raw milk sold directly from farmer to consumer, and there are exceptions for people drinking the raw milk from their own animals. This has given rise to the concept of selling "cow shares" rather than the milk itself. If a customer buys a share in a cow, they are technically drinking milk from their own cow and bypassing the law. This can be a tricky area to get into, as some states will allow this practice without question while others are changing their laws to prohibit cow sharing for this purpose.

If you are going to sell raw milk where it is legal, you will have to make sure your products are all clearly labeled with a warning.

Aside from the raw/pasteurized milk debate, you will need a state dairy license in order to sell milk legally. Contact your state Department of Agriculture for the details on how to apply for the right permit. Only through your own state can you find the right requirements and inspection and grading procedures for your area.

Selling meat

As mentioned in Chapter 1, the actual butchering process is better left to professionals so you do not have to deal with the added inspections and regulations involved in butchering in-house. If butchering is contracted out, you only have to manage the legalities of *selling* the meat. As with milk, getting a meat seller's license is handled at the state level, so you will need to contact the Department of Agriculture in your state to find out the proper requirements and application process for the license.

In many cases, you can bypass the need for a license if you do not actually handle any meat on your farm. By selling a half or quarter of an animal at a time, you technically can consider it a sale of an animal. Your customers then arrange for the animal to be processed at the slaughterhouse, and they pick up the meat themselves. Although this is convenient from a regulation standpoint, you will limit your customer base if you only sell whole animals. For half or quarter sales, you need two or four customers ready to purchase at the same time, depending on how the animal is divided, so that the entire animal is sold with each transaction.

You will have more flexibility if you do have a seller's permit and keep an inventory of smaller cuts of meat for the more typical customer. There is more risk with this approach, as you can end up with unsold cuts of meat that eventually go to waste. Certain cuts are more popular than others, and each animal you slaughter is going to produce the same number of cuts each time, whether you want them or not.

One notable exception for most states is that small-scale meat chicken operations, where chickens are sold for meat rather than the eggs they produce, are not required to get a license to sell meat. If your farm produces fewer than 1,000 chickens each year, you can bypass the permit altogether. This minimum only applies to chickens.

Selling eggs

Again, the regulations involved in selling your own eggs fall under state jurisdiction. The rules for selling eggs legally are much less rigorous than for meat or milk; thus, it is a simpler prod-

uct for your farm to manage. Permits are inexpensive, and many small-scale farmers end up being exempt anyway. In some states, anyone who sells his or her own eggs directly to consumers is exempt from licensing, regardless of the size of the operation. Sometimes, the regulations are based on how many hens you have, with the minimum being 100, 200, or even as high as 500 before you need to get an official permit for sales. In yet another scenario, you will need a permit only if you sell a certain number of eggs; often, the minimum is 30 dozen per week.

If you are exempt from the requirements, you will most likely have to label your eggs as "ungraded" to identify that your eggs have not been inspected, but the exact rules will vary from state to state. Having your eggs graded is a USDA service, or it can be done at the state level, which will result in a state label rather than a label provided by the USDA.

Taxes

Taxes are complicated even at the best of times, and dealing with them as a small-scale farmer is no different. This only applies if you are running a farm for profit. Having livestock for your own personal use should not have much impact on your tax situation.

How you file your taxes will depend on how you have structured your business. As explained earlier in this chapter, some forms of business will require separate tax returns, but if you are running a sole proprietorship, you can add your farm income in with your personal tax returns. In particular, you will have to file a Schedule F, which is an additional form to file with your income tax returns that lists income from self-employment and expenses associated with that business. It is similar to the Schedule C used for other small-business activities, but in a Schedule F, the deductions are itemized for agricultural operations. There are not any specific benefits or breaks from filing a Schedule F; it is just easier to organize your deductions with forms designed for farming.

You might want to have a professional accountant do your taxes, especially during your first year of operation, until you get a better understanding of how it works. During the startup phase, you are going to have more big-ticket purchases, such as vehicles, equipment, and buildings, which will need to be properly depreciated for tax deductions.

In order for your farm to qualify as a true business and not just a glorified hobby, you have to make money or at least show that your intention is to turn a profit. No one expects you to be in the black right from your first day, but at some point, you will have to earn money.

There are more to taxes than annual income tax. Local property taxes will consider agricultural use, providing tax breaks to small farmers. These breaks will vary greatly from area to area, so contact your local county office for more information. In many cases, you will have to have a certain amount of land dedicated to your farm and make enough gross profit per year in order to qualify.

This chapter covers the basics of business and finance, but do not consider this a minor part of your business planning. In order to understand the implications of loans, taxes, and legally selling food products fully, you should consult both a lawyer and a financial adviser before starting your farming business.

CASE STUDY:
START SMALL,
AND BUILD OVER THE YEARS

Bob & Beth Van De Boom
V D B Organic Farms
N5181 Pinnow Rd.
Delavan, WI 53115
www.vdborganicfarms.com

Back in 2002, VDB Organic Farms was just 12 acres of land and 10 sheep. Over the years, they have worked to build up the soil quality, gained their organic certificate, and now have expanded to nearly 120 acres (partly owned and partly rented). Their livestock population is geared towards meat production and includes mainly cattle and a few sheep. They have about 100 cows, calves, as well as steers, on pasture for their beef supply as well as a several hundred chickens to be sold for meat.

Why did they go organic? According to Bob Van De Boom, "We have been involved with the organic foods movement for over 20 years. We purchased our very own farm in 2003 after having little luck finding producers who met our high standards in regard to taste and quality as well as treatment of the land and livestock. We did not want antibiotics, growth hormones, nasty fly sprays, or other chemicals in food that we consumed."

Because Bob already knew quite a bit about the organic industry, he found the process quite smooth. As with most farmers, the biggest difficulty was keeping track of everything. He recommends that you write everything down. VDB Farms is certified with MOSA (Midwest Organic Services Association). He was lucky to find a vet who was already on board with organic ideas, thanks to an organic dairy friend who paved the way for them. They also work with several local processing facilities for slaughter and butchering, rather than doing that work themselves.

VDB Organic Farms has a herd of Murray Grey cattle, St. Croix hair sheep, a mix of laying hens and Cornish Cross chickens for meat. All of their beef, lamb, and poultry are grass fed and finished. They make good use of the Internet and their own website to advertise their wares for direct-to-customer sales.

Bob's biggest regret would be the purchasing of haying equipment in the hopes of doing that job themselves. He has since learned that it would have made more sense to buy their hay instead of taking the time to harvest it himself. Not only that, but you are bringing added nutrients *into* your farm by buying hay.

They do not intend to expand much further, and Bob said starting small is the best way to go. "It is better to realize you do not like what you are doing when you are small than when you have your whole life and savings tied up in to it and realized you made a mistake. Do not do it for the money; do it because you love doing it."

Chapter 3

Growing Crops and Forage

ot all animal feed must be trucked in from a store. There is no reason why you cannot grow some of your own. Although you will not be able to produce the specialized feed pellets that work well for some livestock, you can certainly produce your own basic grasses or grains for feed.

Why Grow Your Own Crops?

There are several reasons why you might want to grow your own crops for animal feed. The one that usually comes to mind first is the cost. Rather than paying top-dollar for organic feed, you can save a lot of money by growing it yourself. There will be added costs at the outset, possible new equipment if you want to plant and harvest large areas, and new outbuildings to store your feed, but eventually, the overall cost will come down.

For the organic farmer, the added benefit of growing your own animal feed is you can trust that it completely complies with organic standards. If you are unable to find a feed dealer who offers a decent selection of organic animal feeds within delivery distance, this might end up being your only option.

If you do well at growing crops, you might even find that this can become an additional income source for you. Natural and organic feeds are gaining in popularity, so there is usually demand. Some grains, such as oats or barley, can even be used or sold for human consumption as well as for animal feed.

It might seem practical to grow your own crops, but the time, space, and energy might not make it a suitable project for all small-scale farmers. It can pull your farm in two separate directions: trying to manage herds of animals as well as fields of growing crops. For someone just starting out, this can turn your farm into an unmanageable project. Even if you have the space to grow crops, wait a year or two before diving in. Get comfortable with your animals first.

Choosing Which Crops

How you choose which crops to grow will depend on the needs of your animals, how much space you have to dedicate to planting, and which crops will grow in your climate. Hay is the most common and can be grown almost anywhere for any animal.

After hay, your next potential food crops will be grains. Oats, barley, rye, and wheat are all good food choices, though none are used exclusively. Do not forget vegetables. They can be grown just as easily for animals as they are for humans.

Hay

In almost every case, hay will be an important food source for your animals, and it is a good place to start when considering potential food crops. Hay is grass that has grown to maturity, been mowed and allowed to dry, and bundled in bales. The type of grass used for hay does not matter, so you can usually grow some type of potential hay in any climate.

If you want to go the easy route, you can just let a field grow wild and mow down the natural grasses. However, the nutritional content will be better if you plant a field with good-quality grasses that will create a solid diet for your animals. Timothy and alfalfa are two excellent choices.

Hay is a relatively easy crop to grow, but the machinery needed to mow and bale the hay might not be practical. Contracting out the harvesting of your grass is one way to get around the equipment needs, and it will save you some time, too.

After the hay is cut and baled, you will need to store it. Compared to feed grains, hay takes up a huge amount of space. If you had already planned to accommodate the size when buying hay from someone else, you only need to be prepared to store an entire harvest at once. You will not be able to grow hay just four or five bales at a time, unlike buying four or five bales from a local hay supplier.

Grains

In the grain category, you have several choices that make good food sources for most livestock animals. The typical choices are oats, barley, wheat, and corn. Some animals will do well with soybeans, too.

Once harvested, dry grains store well, so they are excellent winter foods. Unlike hay, they take up little space; yet have high caloric content for their volume. They are high in fat compared to grass or hay, which is why they should not be the main component of any diet.

The downsides are that you will need a large plot of land for growing grains and that the harvesting can be time-consuming. If you do not have enough space to grow your own, you can always grow a small patch of oats or wheat and use your homegrown grains as a treat or to offset the cost of your purchased grain feed.

Vegetables

Even vegetables that would be fine on your own table can be grown with livestock in mind. They are not always thought of as an alternative to more conventional animal feed, but they should not be ignored. Vegetables are not necessarily easier to grow than grains, but they are more familiar to most people who are used to gardening. Check on the nutritional requirements of your animals before going this route, as not all vegetables will be suitable for all animals. Root vegetables such as turnips are particularly popular for animal feed because they store well and are easy to grow. *The specific animal chapters later in this book will provide more on each animal's needs.*

Turnips

In large quantities, vegetables can be cumbersome to harvest and might not store well for long-term feeding plans, such as over the winter. However, they make a great addition to your animals' diets, both for added nutrition and to give them some variety.

You do not have to grow vegetables solely for your animals, either. Feeding them scraps, cuttings, peelings, greens, or damaged vegetables from your own garden gives them some variety and added nutrition without needing to dedicate a whole new garden just for your livestock. However, doing so would mean your vegetable patch must be kept organic. Anything growing nonorganically cannot be used as animal feed, even casually.

Soils and Planting Basics

Not all crops have the same requirements, but you will want to have reasonably fertile soil in order to do any large-scale gardening for livestock. Soil heavy with clay or rocks or too light with sand will not grow crops successfully.

Vegetables will need the most fertile and loose soils, whereas grains can handle themselves better on less pampered ground. Either way, good, aged manure will do wonders for your harvest. Because you cannot use synthetic fertilizers, manure will be the best way to add nutrients to the soil. Spreading a layer of manure over a season or two can drastically improve any field, so you might be able to grow some of your own crops after a few years of keeping animals if you naturally fertilize in this manner.

Planting grain involves little other than spreading the seed. No digging or measuring is necessary, though the small seeds should be spread out evenly. Birds can make off with a lot of your seed if you just leave it to sit on the top of the soil when you are done. Run a rake over the area to lightly cover up the seeds. Use the same procedure if you are seeding out a field for any type of hay.

A vegetable garden will take considerably more work. After deciding which vegetables are best for your animals, you will need to establish how much space each plant will need and plant your seeds accordingly. Tall, thin stalks of grain can thrive with virtually no space between plants, but vegetables need room to grow or you will get a stunted crop at harvest time. You also need to take care to plant your seeds at the right depth. Some seeds should be just under a cover of soil, and some need to be planted up to two inches deep.

Planting equipment

The equipment required to plant vegetables is minimal. Small shovels, trowels, a hoe, and a rake should suffice to help you get the seeds in the ground. Stakes and string can help you mark off your rows to keep track of where you planted. This can be helpful during the first few weeks when the new seedlings look a lot like possible weeds.

For grain and hay crops, you might want a few more items to help speed up the process. Planting large numbers of seeds can be done by hand if you want to take the time to do so. A broad-

cast seeder will be handy. Small ones can be carried or pushed on wheels, and larger units can be pulled behind a small tractor. They save wear and tear on your arms and will sow the seeds more evenly than you can by hand.

Harvesting

Planting grains and hay can be straightforward and simple, but harvesting is where the work comes in.

Harvesting hay really just means mowing it all down. The entire stalk is used as food, so you do not need to do too much work to process your hay. Cut it, and then leave it out in the sun to dry. Plan to do this chore when there is no rain in the forecast for a few days. Depending on the temperature and amount of sunshine in your area, it can take as little as two or three days or up to a week to dry.

The next step is the most difficult one. You need a way to bundle it into bales. Small balers that pull behind tractors can be used and can be a good investment if you plan to grow your own hay each year. Otherwise, you will want to hire someone to do it for you.

Grains are a little more involved. When the seed heads have matured and completely dried on the stalk, it is time to harvest. The stalks will need to be cut, and the grains shaken loose. This is called threshing and can be done quickly with the right machinery or done more slowly by hand. Once the grains have been separated, you can store them as you would any purchased grain. With some animals, you might not even have to go through the threshing stage. Just mow down the plants and store them intact as you would store hay.

For livestock vegetables, you will have to wait until your specific plants have matured and pick them by hand. Machinery is not helpful in this case. Once harvested, store your vegetables somewhere cool; the specific storing instructions will depend on what you are growing. A cold root cellar environment is best for many crops and can help extend freshness.

Grass and Grazing

Until now, the theme has been to grow a crop, then harvest it to feed your animals later. This is not your only option for growing your own feed, however. Grass feeding your animals on pasture is an increasingly popular choice.

With grass feeding, you will not have to deliberately go out and feed your animals, as they are left on their own to eat as they graze around your field. Although it is the simpler and healthier choice, it will require a little planning in order to make it work.

First, you need to have enough space. Planting an acre of oats will give you an acre of oats at harvest time because the entire space is used for that one purpose. However, growing food on an acre that has large animals roaming around on it will not produce the same acre's worth of food. Therefore, plan on using more space for pastures than you would with a strictly hay diet. Exact requirements will vary by animal, but a cow needs three to five acres, and you can keep a goat on less than half of that space.

You will have to reseed your pasture to keep it continually growing the right foods for your animals. Do not let it go wild after a few seasons. The need for this depends greatly on your livestock though. Goats will be happy to graze on anything that grows, but cows need a more consistent diet. You will have to manage your fields in accordance to your animals' needs. If you can manage the space, maintain at least two fields large enough for your herd. Have one seeded and growing while the animals graze in the other one. Then, switch your animals to the other field when the new plant growth is high enough.

Allowing your animals to graze naturally for most of their food is great, but usually, it will not work as a year-round choice unless you live in a mild climate that allows for year-round growing.

You will likely still need to have a source of food available during the winter and early spring months.

No matter which plants you grow for your animals, you cannot use any artificial chemicals on them. You will have to avoid all synthetic fertilizers and pesticides, or you will no longer be raising organic animals. With the manure left behind as they graze, fertilizing a pasture is seldom going to be necessary anyway.

Growing your own food can be a large endeavor and is not necessarily a task you want to begin immediately. If you have the space to do so, it can create a large cost savings for your farm as well as provide a healthier natural food source for your organic herds.

Animal Husbandry

Some small-scale farmers keep their operations simple by bypassing the breeding component of raising animals. They buy young animals to raise to slaughtering age and then buy new animals the next year. With the added costs of young animals, such a system will not net you as much profit, but it does make life a little easier if you do not want to handle the organization needed to breed your own successfully.

The animals you are raising and what their purposes are on your farm will determine your breeding needs. A sheep farm that only uses its animals for wool will not have as much need for new lambs as one that raises sheep for meat. Any time animals are not slaughtered, you will have a less pressing need to keep your stock replenished. One exception is when raising animals for milk. Although you do not slaughter milk goats, sheep, or cows, they need to be regularly impregnated in order to keep producing milk.

Breeding Basics

It might seem daunting to manage a breeding program for a herd of animals, especially if you have no knowledge of the basic concepts. However, genetics is actually quite easy to understand and will help you start a successful breeding program.

The overall principle you want to work toward is the breeding of two animals with good traits that are not too closely related to each other. If you stick to these simple concepts, you are well on your way.

The simplest approach is just breeding animals in order to replenish your herd, with no consideration to any particular traits. This approach eliminates the need to understand genetics, but most farmers who breed their animals usually do at least a little selective breeding.

Genetics

Genetics might make you think back to your high school biology class, which might not be a happy association. It can be a complex subject, no matter how you look at it. The amount of genetics you need to know for your farm will depend on your breeding purposes. Are you just looking to keep your herd producing offspring, or do you want to breed certain traits in your animals? Are you breeding for simple traits or more complex ones?

For simple breeding programs, this chapter should be sufficient. But if you intend to establish a more detailed program that spans over many generations in order to produce certain results, you will need more advanced genetics study.

The genes within DNA control all traits, but the traits we see are only half the story. Because all animals get their genetic information from both a mother and a father, there are two copies of each gene for any one given trait. One can be dominant (what we see) and one can be recessive (what we might not see). A look at eye color will illustrate this point.

In this example, brown is the dominant color and blue is the recessive color, which means a person with a brown gene and a blue gene will actually have brown eyes. Two brown genes will produce brown eyes, and two blue genes will produce blue eyes. This seems simple enough, but it means you cannot necessarily judge genetics by what you see. A person with brown eyes might have a blue-eyed gene as well. And when it comes to that person's offspring, it will make a difference.

To move this to a more topical example, consider the wool color of a sheep. If you are looking to breed your sheep to create more brown-fleeced animals, it might not be as simple as breeding brown males with brown females because unseen recessive genes might be involved, which can create white-fleeced offspring. This is a hypothetical example and not a true genetic issue with sheep fleece. In order to establish which genes are present, you will need to keep records about offspring and their parentage.

To continue with this example, a mating of two brown sheep could produce either all brown offspring or lambs with white fleece. Having white offspring would mean that both parents had a recessive white gene in hiding. Because brown is the dominant color, it is never the hidden gene. White-fleeced lambs only carry white genes and should not be bred again for more brown sheep. The same is true for the brown parents, now that you see they are carrying white genes. You can certainly breed the parents again if you wish, but you will be continuing the line of white genes in their offspring by doing so.

It might seem difficult to grasp — and even more difficult to envision how you would plan to account for such things. However, after a few seasons of breeding, you will see patterns emerging in your records that will be much clearer to understand. At this point, this is mainly just genetic theory. *Actual techniques and practices are covered in the next section of this chapter.*

Of course, these are just simplified and hypothetical examples. Even with eye color, there are many variations in the real world. Eyes can actually be green, grey, hazel, or even violet. It is never just brown and blue. More complicated features, such as milk production or muscle building, are usually controlled by many genes, which make them difficult to track. Less tangible features, such as temperament or behavior, are even more obscure and are not likely to be traceable using this type of dominant-recessive gene concept. You will need to keep detailed records and a close eye on your animals to follow the traits you want.

Breeding Practices

Such genetic concepts are abstract at the best of times, and what you really need to know is how to put that information to use on your farm. With a few good techniques, you will be able to master the genetic makeup of your animal herds and have strong offspring every year.

Your first step will be determining what information is already known about your animals' genetic makeup. Before you can plan your breeding goals and find the best techniques, you should be armed with all the right information. Because livestock have been rigorously bred for various traits for so many generations, a great deal of information is already known. Use that as your starting point. This is specialized information, but your state's livestock associations should be able to help you find what you need.

Again using fleece color as an example, most color variations have been thoroughly studied, so you can start by already knowing which combinations of genes produce which colors and which ones are dominant or recessive. Color tends to be the most studied because it is easy to see and measure. The specific genes that control milk production or muscle mass are far less obvious.

The most important aspect of your breeding program, no matter what your goals are, is recordkeeping. You must be able to identify each of your animals. Numbers usually work best, though you can certainly name them if you prefer. Keep records on each one, including parentage if it was born on your farm.

If nothing else, this will allow you to prevent any breeding between closely related animals. Tattoos, tags, or collars are good tools for individual identification. Tattoos are permanent, but tags are more visible and allow you to implement a color-coding scheme to aid in herd organization. Tags also can be changed periodically if you decide to try a different system once you get started.

In order to keep your breeding under control, you will want to keep the males and females separate, except when you intend them to breed. Leaving all of your animals in one pasture and just letting nature take its course is not ideal because you can get undesirable inbreeding if siblings mate. Even if you do not plan to control the genetic traits of your animals, this is still the responsible approach to take. Also, knowing roughly when a female got pregnant will mean you can predict when the birth will take place. If your females get pregnant at random times, births might take you by surprise.

To begin breeding, you will need to have a selection of females and males that are mature enough. Breeding ages will vary from species to species. Cows can be bred at about 18 months, but you can breed sheep under a year old. Meat operations will not find this to be quite as problematic as a farm with a dairy. Milking farms have almost no need for males, whether they are raising cows, goats, or sheep. You can either keep a few males solely for breeding purposes or "rent" male animals when you need them. Keep in mind that adult males of some species are not friendly with each other, and trying to house, even temporarily, a group of males together can result in fighting and injury. If your operation does not warrant raising male animals, there is always artificial insemination. *There is more on this option later in this chapter.*

Unwanted male offspring are another aspect to consider before you begin your breeding program. Farms that raise livestock for meat will use male and female animals, but if you are raising female animals for milking purposes only, males that are born will not be needed. The one exception is that you might want to have a few males around for breeding, which usually does not account for all the male babies you end up with. Figure out your plan for these offspring before you are faced with a stall full of little animals. It is common to have them slaughtered for meat or sold as potential breeding males to other farmers.

You will need to keep extensive records once you begin breeding your animals. This is where many small farmers can get overwhelmed, particularly if you are managing a large number of animals. Each animal must be identified with a name or number, and you have to record each breeding with dates. Then, extend those records when the offspring are born. Record all the traits you are tracking, not just at birth but also as the animal matures. You will not know which goats are the best milk producers until they are older, for example. Also, record their temperament and general health features as they mature. Animals that tend to get ill more often than others should not be bred, nor should any that have a bad attitude toward people.

Does this sound like more than you can handle? Several different software programs on the market can simplify recordkeeping and guide you in creating the best matches for your animals. Ranch Manager by Lion Edge Technologies is a popular software package that can handle a number of types of livestock and allows you to manage breeding and health records for all your animals. A simple three-ring binder also can work well if you prefer a more traditional route.

Artificial insemination

On an organic farm, artificial insemination (AI) is an acceptable procedure because your animals are not being treated with any prohibited chemicals. It might seem unnatural, but it does fall well within the guidelines of organic farming. There is one twist you should be aware of. Sometimes, semen may be treated with antibiotics to prevent the spread of disease. This is one of the few areas where you do not need to be warned off the antibiotic usage. Using semen treated with antibiotics for AI will not be a threat to any organic certification because it does not have any effect on your animals.

Large animal vets can either perform this service for you or help you find someone else who can. You also can save a little money  if you order the semen on its own and perform the procedure your-self instead of hir-ing a technician to do it. With a little experience, this is definitely a viable option. The sim-plest method is to insert a quantity of semen into the vagina of the cow, but you will get better impregnation results with a new method that includes a catheter to deposit the semen directly into the uterus.

Specific costs to have AI done for your animals will vary by re-gion and by the species of animal you are working with. For large animals such as cattle, it will cost roughly $30 to $50 per insemi-nation, which does make the process much more cost-effective than keeping and maintaining a bull to do the job for you. This is particularly the case with a smaller farm where pasture space is at a premium.

When using AI, you will still need to be aware of when your fe-male animals are in heat because that is still the only time when they can become pregnant.

Another benefit of using AI is that you can use sperm from any male animal, regardless of geography. This allows you to be more creative and selective in your breeding program because you can breed your females with other breeds from anywhere in the world. AI will help maintain a purebred line or allow you to try cross-breeding with animals from other regions.

There is at least one area where you will have to forgo the benefits of AI, and that is in some cases of purebred horse breeding. Certain breed standards will not allow for AI to be used. Only a natural mating between a mare and stallion are allowed in the thoroughbred breed if you want the offspring to be registered, for example.

Facilities

You will need to have pasture and stalls arranged so you can keep your adult males and females separate. If you are not keeping males on your farm, this might not be necessary. You will also need to have separate space for birthing and living space for mother animals and their newborns. This does not have to be two areas. Unless you plan to have many pregnant animals at once, the stall used for the birth can also be used to house the animals afterward. Pregnant animals can be housed with the rest of your herd until they give birth.

Either way, it should be a clean area of the barn, well secured, and free from clutter. When dealing with a birthing animal, you do not want to accidentally trip over a bucket. Under normal circumstances, it is easy to remember there is a sack of feed inside the door or that the broom leans into the walkway. When an animal is giving birth, your attention will be focused on the animal alone, so keeping the area free of potential hazards will prevent accidents. It should be well lit without being overly bright. Other than these requirements, you will not need too many other specific facilities to manage your breeding efforts.

Unfortunately, animals do not always go into labor right on schedule, so you might not always have the opportunity to isolate your females before they have their babies. In that case, bring them to your newborn area and make sure the baby is healthy.

Even if you feel you are fully equipped to handle the birth of an animal on your farm, have your veterinarian present for the first season or two. The birthing process is completely natural and should not require intervention, but it is prudent to be ready for emergencies for the health and safety of your animals. Once you have some experience, you can handle it on your own, but keep the veterinarian's number handy anyway. Complications can happen quickly.

The list of necessary supplies can be extensive if you intend on handling a birth on your own, and the expertise needed to use them is beyond the scope of this book. Basic supplies would include iodine, latex gloves, old blankets and towels, syringes, soap and water, obstetric lubricant, flashlights, and scissors. What counts more is the knowledge to use these things; so further research is necessary before you can safely assist in a birth without professional assistance.

Reproductive Health

Nature is great at handling pregnancy and birth, but a prudent farmer makes sure to take every precaution to care for breeding animals and their offspring. Having a vet handy is the most important step, although knowing a little about prenatal care yourself can go a long way in keeping your animals healthy.

Humane practices

As an organic farmer, not only do you want to keep all artificial chemicals away from your animals, but also you want to treat them ethically and humanely. This can impact your breeding plans, as it is important to not abuse your animals with excessive breeding. Just because a female animal is capable of being pregnant again does not mean you should rush to rebreed her immediately. Give your livestock a chance to recover from birth before starting the cycle again.

Also, it is humane to allow the mother and her offspring to spend time together. Just because you want to start milking the female does not mean she needs to have her baby taken away. You might get a lower yield, but she will definitely produce enough milk for both of you.

Feeding and feeds

Pregnant animals might need additional feed or nutritional supplements, depending on the specific animals you are raising and what their regular diet consists of. Talk it over with your veterinarian to make sure you are providing the right feed. Once the young are born, you might also have to adjust the feed again to help compensate for all nutrients going into the milk the mother is producing. Additional calcium is one common supplement for nursing animals. Nutritional supplements will not interfere with your organic goals as long as they do not contain synthetic materials.

Again, keep your organic goals in mind. Double-check that all feed is organic and that there are no hidden antibiotics or medical additions to the feed. The risk is always there with regular feed, but when you are buying special feed for nursing females or young animals, it is even more difficult to find antibiotic-free products.

If the mother animal is unable or unwilling to nurse her young, you will have to bottle-feed. Milk replacers are common on the market and should not be difficult to acquire, including several brands with organic ingredients. Many are species-specific; so make sure to buy the right replacer for your animals. Calf's milk replacer will have different contents than goat's milk replacer, and you can give a newborn animal digestive problems if you use the wrong type. Some generic multi-species formulas can be handy to keep around, just in case.

Bottle-feeding will take some dedication and a considerable amount of time. The volume of milk and the frequency of feeding will vary from animal to animal. You will need some bottles and appropriately sized nipples as well.

Once your young animals are beyond the newborn stage, they typically eat the same foods as the adults. But if you are offering additional grain or more concentrated food to young animals, you will have the challenge of keeping the adults out of it. A creep feeder is the answer to this problem. Build a small area for a feeder, but add a barrier around it that will allow small animals through while still keeping out the large ones. A bar or board horizontally placed across the entrance to the feeder would let calves or kids duck underneath while keeping out the adult animals. Another variation would include vertical boards or bars placed too close for adults to squeeze through.

Health Basics

Because you are operating an organic farm, you need to focus more on the health of your animals than a conventional farmer might. There are several ways you can improve the health of your livestock without resorting to nonorganic medications and treatments.

Finding large animal veterinarians

In completely rural areas, finding a veterinarian that specializes in large animals such as livestock usually is not difficult. However, if you are starting a small-scale farm on a smaller parcel of land in a more urban setting, you might have a challenge ahead of you. A regular veterinarian will not be able to help you.

Like getting good contractors, the best way to find trustworthy veterinarians is by word of mouth. Ask other local farmers which veterinarian they use, and get some referrals.

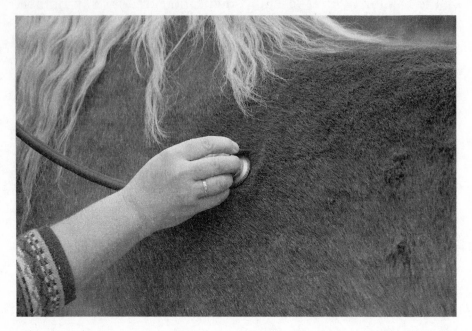

Before you hire a veterinarian, find out his or her views on organic practices. Trying to work with a veterinarian who does not have the same viewpoint you do will be difficult because he or she will not be open to nonchemical treatments, particularly the avoidance of antibiotics. An "organic-friendly" professional would be best, but if you do not have the luxury of a choice, working with a conventional veterinarian will have to do.

Because livestock animals are not easily transported into the vet's office, it is typical for veterinarians to make house calls. Ask about their on-call policies. Most will come out any time you have an emergency, but some only offer their services during specific business hours.

Preventive medicine

Preventive medicine is more important with organic farmers than conventional ones because a standard treatment of antibiotics is not an option. You need to keep your animals from getting sick in the first place because it will be harder to treat them once they are ill and retain their organic status.

Vaccinations are one aspect of preventive medicine that is fully available to an organic farmer. Under the USDA rules for organic livestock, vaccinations are allowed without any detriment to organic labeling or certification. This includes both synthetic and non-synthetic vaccines. Therefore, you can comfortably vaccinate your animals without any worry about future organic status. In fact, it is recommended that vaccinations be done to help ensure good health and well-being for the animal later in life, which is part of the organic goal.

Vaccinations take care of many potential diseases, but animals also are prone to worms and parasites that also require steps to prevent. Traditional wormers are not an option under an organic system, so you will need to be nonconventional. Two of the best all-natural options are garlic and diatomaceous earth (DE).

Garlic is a natural antibacterial agent, and the taste of garlic is strong enough to make your animals likely to turn up their noses, so do not expect them to willingly chew on a few cloves. Puree your garlic, and be prepared to force-feed with a large livestock syringe.

Garlic will not have any impact on your meat or eggs, but too much of it can affect the taste of your milk. Whether you are milking cows, goats, or sheep, a lot of garlic in their diet will come out in the milk. Thankfully, a worming treatment of garlic is not an everyday chore so you just have to remember to dispose of your milk a few times a year after you have treated your animals. The exact schedule will vary by animal species and size, as well as what the local parasite population is like. Your vet should be able to give you specifics, but usually it is done only three or four times each year.

You are surely familiar with garlic, but DE might be more of a mystery. It looks like a fine white power, much like baby powder. It is *completely* harmless to the touch and does not pose any health risks to people or animals of any size. However, on a microscopic level, it is made up of tiny sharp pieces of diatom shells. To an insect or parasitic worm, DE is like a dose of broken glass. It quickly cuts them up and kills them. Food grade DE can be mixed in with your livestock grain, both to keep out insect pests that would consume the grain and as a wormer after it is eaten.

Unlike stronger chemical wormers that can be administered once or twice a year, you will need to dose your livestock more often when you use these natural methods. Regardless of the specific remedy you choose, talk to your veterinarian first. Ideally, he or she will be sympathetic toward organic goals and will be able to provide advice and dosages for you.

Many injuries and diseases can be traced back to unclean living conditions and lack of space for each animal. By following a more ethical approach to raising animals, you are already preventing many possible health hazards. A clean farm and plenty of room will go a long way to prevent illness.

When all is said and done, the well-being of the animal should be of high importance. If antibiotics or other synthetic medicines do become necessary, an organic farmer should never let an animal suffer in order to keep retain an animal's organic status. Although treating a sick animal might take it out of the running for an organic label, its meat is certainly still fine for consumption and can be eaten by your own family or sold as nonorganic.

Keeping a clean farm

Nothing helps keep animals healthy like a clean farm. Of course, the term "clean" has a completely different definition when talking about a barn or pasture compared to the inside of your house.

When you think about keeping your farm clean, do not worry quite so much about actual dirt. The biggest problem in cleanliness is manure and other animal waste.

Small pasture areas will soon become overwhelmed with manure, but larger grazing areas can usually handle their manure

load without becoming overly soiled. It breaks down well and eventually "disappears" to become part of the soil. If that takes too long and your animals frequently have to walk in their own droppings, you will have to deal with it before health problems arise.

A small tractor or truck with a plow on the front can scrape a pasture clear of manure in short order, though doing so by hand with shovels might be reasonable in a small enough space.

The next place to watch for cleanliness is inside the barn. Because your animals are in a smaller space while indoors, waste will build up much faster than out in the field. Keeping the stalls clean will be a regular chore — and not one to be shirked. The specific schedule will depend on the size of your barn, the number of animals you have, and the sizes of those animals. Large animals in small spaces likely mean you should shovel out the stalls every other day or even daily. You might not need to scour the stall right down to the floor that often, but remove any obviously wet straw and piles of manure.

Food troughs and water buckets need to be washed out regularly and treated with a disinfectant periodically. There are organic disinfectants, so you do not have to worry about contaminating the water supply or food with prohibited chemicals. Even so, all containers should be well rinsed after cleaning.

Some animals habitually will knock their bedding into their food or water areas, so you might have to hang the containers from the wall or use some other trick to keep them up out of the bedding.

Your own animals are not the only sources of dirt on a farm. Even the most tightly built barn can have a mouse or two at some point. Rodent droppings can spread disease, so regularly check the corners and behind large items for signs of mice. A good cat on the premises can help, and strategically placed traps will keep the pest population under control.

Once a year, clean out the stalls completely, and then wash them with disinfectants. It is a good spring chore after the long winter of increased enclosure.

CASE STUDY: SOME FARMERS PREFER TO STAY SMALL

Mary & Jim Moran
High Ridge Meadows Farm
1800 Chelsea Mountain Road
East Randolph, VT 05041-0125
http://highridgemeadowsfarm.com

High Ridge Meadows is primarily a meat farm; farmers raise cattle and Icelandic sheep for beef and lamb. They also take advantage of their sheep's fleece to sell yarn, rovings, and even finished knitted products. Add some chickens and a few ducks, and that rounds out the livestock population of the farm. Future goals for the farm include a few pigs for pork and letting the cattle herd grow. Mary Moran said she prefers the smaller farm because it allows her to maintain the highest quality for her animals.

The farm currently has about a dozen each of cows and sheep and a flock of 100 chickens that all live out on pasture. They began with just two head of cattle when they first ventured into keeping livestock. The farm also has an orchard, vegetable garden, and herb garden. They have been certified organic since 2005 and found the process daunting but not difficult.

Moran said there was some confusion during certification about their hayfield, and she had to manage a great deal of paperwork through it all. But because she already followed organic principles, there was little to change around the farm. With a hayfield that is certified organic, they get most of their animal feed from the premise, and there are other organic suppliers nearby for other feed. Herbs for the sheep are available for online purchase.

They work with an organic-friendly vet who is happy to oblige with natural and organic treatments for their animals. They also use other outside assistance for many chores and tasks around the farm as well, such as for plowing, tilling, and splitting wood. A couple of neighborhood children help keep an eye on the animals when no one is home.

Sales for High Ridge Meadows are primarily wholesale. They sell their products to various markets and restaurants in Vermont as well as in Massachusetts.

Moran has a few suggestions for the novice farmer. "Start small and take it slowly," she said. "Have your infrastructure in place before taking on anything new. It is amazing how all the planning in the world can go wrong. Nothing ever goes as planned, so expect the unexpected. Do not believe everything you read."

Chapter 5

Pigs

> **Learn the Lingo**
>
> - **Shoat:** A young pig, usually still unweaned; also called a piglet
>
> - **Boar:** A male pig that has matured to breeding age
>
> - **Barrow:** A castrated male; also sometimes called a stag if the castration takes place after puberty
>
> - **Gilt:** A female pig young enough to have not yet been bred
>
> - **Sow:** A female pig that has had her first litter
>
> - **Farrow:** To give birth

Pigs are one of the few livestock animals that will really only provide you with one product, which is meat. It makes them less flexible but also simplifies any farming program.

Pros and Cons of Raising Pigs

Raising pigs will mean you are able to sell well-known meat cuts, and you will not need to convince customers to try anything unusual with your products. Ham, bacon, and pork chops will need no explanation.

One major advantage to raising pigs is you can have an animal large enough for slaughter in only six months that still produces a worthwhile amount of meat. You can easily get 150 pounds of pork from a 6-month-old pig. This allows small-scale farmers to eliminate some chores through the winter if they wish by purchasing young pigs in the spring and slaughtering them in the fall of the same year.

A popular misconception about pigs is that they are dirty animals to keep. In reality, they are intelligent, and unless you house them in tiny enclosures, they are clean animals. Of course, this does not mean they are perfect animals. Pigs have an instinctive need to dig and root, which can cause havoc with fences and your pasture. Their low bodies can make them difficult to catch and handle, though they are not generally hard to manage.

Housing and Pasture

Although pigs do not give the impression of being a grazing animal like goats or cows, they will do well if kept on pasture to forage for themselves in the same way.

You will not need to build high fences to keep your pigs enclosed as you would with animals such as goats. Pigs will not leap or climb over fences unless you neglect them, and they are in search of food. A standard fence made from rolls of woven wire or even a wooden slat fence will only need to be about 4 feet tall.

One thing you do need to keep in mind is a pig's natural inclination to root or dig in the dirt. They can do a lot more damage to a pastured area than other livestock animals, especially if the field is small. This can also impact your fencing plans. Any traditional fencing will need to extend below the surface of the ground or your pigs might end up digging around the base of the fence and squirming underneath. If you bury the fencing 1 to 2 feet below the surface, you should have little trouble. Another option is to run a line of electric fencing close to the ground or just use electric fencing completely.

Electric fencing is particularly well suited for pigs because they are not prone to jumping, which means you can use a fence line of only one or two low strands of hot wire, rather than maintain a 6-foot-high fence. Always try to have one line at "snout height," which can be adjusted as your pigs grow. Due to their intelligence, most pigs will touch a hot wire once and learn the lesson immediately. With wire or wooden fencing, you will only need to build it to a height of 3 to 4 feet.

If you are not allowing your pigs to roam around on pasture, your outdoor pigpen should have at least 200 square feet of space per pig, though more is even better.

For housing, pigs are not needy animals. As mentioned earlier, you can buy and slaughter pigs in the same year and avoid the need for winter housing. If you are going this route, your pig enclosures can be simple because you do not have to deal with keeping the building winter ready. Inside, a layer of straw makes for good bedding that is easy to clean when necessary. Pigs usually stay away from their sleeping areas when going to the bathroom, so they might not make much mess within the shelter.

Because of pigs' low stature, a pig house does not need to be tall, though if you build it too low, you might have trouble getting inside to clean it.

Although you can harvest your pigs within a year, you do not *have* to take that approach. There is no reason why you cannot maintain a breeding herd over the winter if you do not want to buy new shoats, or young pigs, each year. If you are going to overwinter your pigs, you must make sure their home is draft-free but still ventilated. For pigs that will stay indoors for most of the winter, plan on a house with 150 square feet per pig. You can keep them in closer quarters if they will also have access to the outdoors for some of the day.

Of course, if you are keeping pigs along with other livestock, you can also keep your pigs in a central barn with the rest of your animals.

Pigs can suffer in hot temperatures; so whichever way you arrange their living arrangements, they must have access to shade during the summer. A place in their pen where you can add water to create a mud wallow is another idea that will add to your pigs' summer comfort. Any area where water pools and has exposed dirt will do. Pigs will naturally root around in such a wet spot, creating a deep wall of mud. Mud is also helpful for keeping the insects away, as pigs do not have fur like most livestock animals do.

Breeds

Many breeds of pigs are used in meat production, though dealing in 100 percent purebred animals is not usually a typical practice. Most pigs raised for food are mixed breeds or crosses of some type.

Choosing a breed for your farm will depend on your specific set of circumstances. With pigs, the differences from one breed to the next are minimal, so you might simply be able to take advantage of whatever shoats are available in your area.

American Landrace

The American Landrace is descended from the Danish Landrace and has become a popular breed in the United States. They are typically white in color with floppy ears and a longer body than some other breeds. Their bodies are classed between medium and large in build, so you can expect a good-sized pig once grown to slaughtering size. The term "Landrace" is common in pig breeds, used with the country of origin for the breed. Other examples are the German Landrace or the Norwegian Landrace. They are still considered distinct breeds, though.

Duroc

Duroc pigs also have the same floppy ears as the American Landrace, but their rusty reddish color sets them apart. This active breed can sometimes border on aggressive. When breeding, they often produce large litters and they grow quickly. Durocs are a good choice if you intend to slaughter your pigs after just one summer of growing.

Tamworth

Tamworths are red like durocs but usually lighter in color with a rusty tan shade. Their body shapes tend to be on the lean side, and their meat is lower in fat than many other breeds, which makes them a good meat animal. They are not as common in the pork industry as others are, though.

Berkshire

If you want to try something a little different, Berkshires are gaining in popularity with small-scale farmers, though they are seldom found in large corporate farms. Their meat tends to have a higher fat content but is known for being tasty. These black pigs often have white feet and white patches on their faces.

Hampshire

These are one of the most popular breeds in the United States and are recognizable due to their black and white coloring. They are mostly black with a white "saddle" across their shoulders.

A Hampshire pig will put on a lot of muscle weight but can grow a little slowly compared to other breeds.

Poland China

This breed grows large and will produce a lot of meat per pig. They are black with white feet and look a lot like the Berkshires, from which the breed originated. Polands will have large litters of piglets but the large size of the sows can make it easy for the mother to accidentally crush her young. Extra care should be taken when breeding these pigs.

American Yorkshire

 Usually just called the Yorkshire, these are popular pigs all across the country. They are the classic pink color that most people think of first when thinking about pigs, though they do come in white and spotted as well. Yorkshires will be well muscled and produce lean meat. This breed is also known as the Large White pig.

Red Wattle

The Red Wattle is a lesser-known breed of pig that will produce lean meat. During the era when customers wanted more fat (lard) with their pork, the Red Wattle grew out of favor. In today's market for leaner meat, it is starting to make a comeback, though it is still considered an endangered breed. They are reddish in color and have a fleshy wattle hanging from their neck.

Chester White

Often just named the Chester, this breed is common in many parts of the United States, though not as popular as the Duroc or the Hampshire. They are a fast growing and early maturing pig, which is one reason they perform so well in either commercial or small-scale farming operations.

Equipment

Given their high intelligence, pigs can easily learn to drink water from a valve or "hog-nipple," which would allow you to keep a barrel of water available for your pigs without the risk of spillage or the water tipping over. A standard 55-gallon plastic barrel can be used with a hole drilled at the right height and the small valve fitted and sealed in place. This is a common way to provide water to pigs, rather than a large open container for drinking. Most farm supply stores carry them, and they are not expensive.

For food, you will need a trough that is sturdy and cannot be tipped over. This is a common requirement with any livestock animals. Attaching a heavy container to the floor or wall is a good idea.

Because pigs do not have fur to care for, brushes or other grooming tools are not needed.

Feeding

Although pigs have a reputation for eating anything and everything, the reality is that you still have to manage their feed so they have a balanced and healthy diet.

If you are new to raising pigs, organic commercial pig feed is a good place to start, and it can be used exclusively even once you have learned a bit more about your animals. However, for the sake of cost-efficiency, it is better to supplement that with other feeds. Doing so is healthier than manufactured food and adds variety to keep your pigs interested.

On average, you will be feeding each pig between 600 and 1,000 pounds of quality food by the time they reach their butchering

size, which is usually about 200 pounds. You will want to feed your pigs as much as they will eat at each feeding and offer them food twice a day. If they leave food in the trough, reduce the amount you give them so little is wasted. The exact amounts can vary widely, as it depends on the type of food used as well as the age of the pigs.

Grains and feed

Unlike other animals that can subsist solely on plant material, pigs will need a good supply of protein in their diet. Young pigs need 16 percent protein, which will drop to about 12 percent if you keep them to full maturity. Because of this, you will not be able to raise pigs solely on pasture, no matter how much land you have to spare. You can supplement a pastured diet with commercial pig feed or add protein-rich foods of your own to their diet. Soybeans as well as milk, eggs, or meat scraps will help with this. Just make sure that nothing truly spoiled goes to your pigs, or you risk illness in your pigs.

The main component of their diet can be hay or free-range pasture, as long as you are adding the extra protein items. Grains such as corn, barley, oats, or wheat can make up a portion of your pigs' diet, but these grains do not have to be the main source of food. These foods, especially corn, are high in fat and can lead to fattier meat at slaughtering time. Although some fat is necessary for good-tasting meat, excessive fat will make your final product much less appealing to customers.

Other foods that can go to your pigs include nearly any type of food scraps that come out of your kitchen — organic only, of course. Most vegetable peelings are fine if given raw, but if you are feeding your pigs hard vegetables such as whole potatoes, squash, or turnips, cook them first.

Many people who raise only a few pigs will gather food waste from stores or restaurants to help keep costs down, but this will not be an option for the organic farmer. These sources of food will likely not be organic or at least not verifiably organic, which means they are off limits for you.

Breeding

Although many small farmers only raise pigs during the summer months, there is no reason why you cannot maintain a herd of pigs over the winter in order to breed your own young.

Breeding basics

A gilt, or female pig, can be bred once she is roughly 5 months old as long as she has reached 200 pounds in size. They do usually reach this weight by 5 months; so generally, you will not have to worry. Whether you have your own boar or borrow one for his services, your female will need to be in heat when the two are put together. Gilts or sows will only be in heat for one to two days, so it might take a few attempts for a successful mating. They will go into heat every 16 to 24 days, though 20 days is the average cycle time. Technically, you can breed her at the first heat, but it is not recommended. You will get a larger litter by the third or fourth heat.

It can be tricky to tell when your animal is in heat. Her vulva will get noticeably swollen and red, and she will stand still when you

get behind her and press down on her backside. It might take a cycle or two before you can really see the signs.

The best way to determine whether the breeding has been successful is to watch

for her next heat. If she does not go into heat again in about 20 days, she likely is pregnant.

Gestation for a pig is 114 days. Coincidentally, that is exactly three months, three weeks, and three days, which makes it easy to remember. The safest plan for timing your breeding and subsequent farrowing is to have the female impregnated in the fall for a spring birth. Having newborn shoats in the middle of winter is not a good idea. By birthing in the spring, you can raise the young to butchering weight in the fall. You can use the same sow to breed again, or use one of the new gilts to start the next generation.

Birthing

When your pregnant sow is ready to farrow, or give birth, have a warm and dry place to keep her enclosed. Depending on your climate and the time of year, you might want to have a heat lamp in place as well. This is only necessary if the temperature is below 50 degrees Fahrenheit, or 10 degrees Celsius.

Until you have experience with farrowing, try to have a veterinarian on hand when your sow begins to deliver. Pigs generally give birth easily, and it is unlikely that any problems will arise. The entire event is usually over in two to three hours. If she has settled down to farrow but has not produced a piglet in about half an hour or this length of time goes between piglets once she has started, one might be stuck in the birth canal. With lubricant jelly, you should be able to use your hands to assist.

Pigs usually give birth to between eight and 12 piglets. Once all the piglets are born, the afterbirth or placenta will follow. Do not allow the sow to eat the afterbirth, as it could prompt her to eat some of the young as well. You should not need to handle the umbilical cords, as they will come loose on their own. If they are bleeding, you can tie them off with a piece of clean thread.

Health

Although you should always have a large-animal veterinarian on call when raising livestock, there are some health issues you can treat yourself. At the least, you should know what potential health concerns can arise when raising pigs.

Newborn health

Most sows are able to take care of their new piglets without much interference from you. If you are lucky, you will not have to do much until they are ready to be weaned. Otherwise, you might have to be more involved with your new shoats. In particular, you might have to deal with rejected young or a runt that is not getting enough to eat.

Shoats that are not feeding will require a large time commitment from you for their care, but it will be worth the effort if you con-

sider their future value to your farm. Goat's milk can be substituted for sow's milk, or you can use commercial milk replacer for pigs. Many milk replacers are medicated, so search for an organic product. For the first three days, you will have to feed your piglets every two hours, which includes nighttime. You might need a bottle for a day or two, but even young piglets can be taught to drink from a shallow bowl.

After three or four days of frequent feedings, you can space them out to six per day and skip a couple of the late night feedings. Once they are a week old, fewer feedings will suffice. At all times, though, allow them to have as much milk as they will drink. You will have to continue feeding them until the point when they would be weaned naturally from their mother.

Whether you are feeding the piglets or they are nursing from the sow, you can begin to wean them to regular solid food when they are about 3 to 4 weeks old. At this point, feed them grain mixed with milk until they are adjusted. They can be fully weaned off milk by 6 weeks.

Any males in your litter will need to be castrated. It is simple enough to do, and you should be able to do it yourself once you have observed a veterinarian completing the procedure a few times. Piglets usually are castrated from about 10 to 12 days old. Intact males can have off-flavored meat, even if they will be slaughtered at six months old. Castrated males will also be less aggressive when kept in a herd with other pigs. Keeping intact males for breeding will mean you have to maintain separate housing and pasture space for them. Another option is to castrate all males and use an artificial insemination service for your breeding.

Pig diseases

Certain diseases might or might not be prevalent in your area, so there is no way to know for certain which will be a problem for your herd. Watch for some of these potential threats, and contact your veterinarian when you have an ill pig. Remember that vaccinations will not compromise your organic certification and can be a good safeguard against future disease.

Porcine parvovirus

Parvovirus is common and will be found in most herds without any problems. It only affects farrowing, so if you are not breeding your pigs, it will not be much of a concern. The symptoms of a PPV outbreak include an increased number of stillbirths and generally small litters. You should vaccinate any breeding gilts before they become pregnant to prevent PPV from taking hold.

Leptospirosis

This disease can be transferred from your pigs to the humans on your farm, which makes it an important one to know about. Like PPV, it can cause poor birthing, small litters, and aborted fetuses, but you will see other symptoms as well. Younger pigs might be jaundiced, and pigs of any age will go off their feed. Bloody urine is another common symptom. It can spread through urine, so make sure to keep stalls and pens clean and provide fresh bedding frequently. The only treatment is with antibiotics, which will render those pigs nonorganic. Vaccinations are available.

Erysipelas

Erysipelas is another commonly found bacterium that will only cause illness in your pigs in dirty or contaminated surroundings where bacteria can grow in large numbers. In acute cases, your pigs can die with little outward symptoms. A high fever, obvious listlessness, infertility, and skin lesions can also exist. Keep stalls clean, and do not let your pigs spend too much time in their own waste. There are vaccinations for this illness, though you might have to vaccinate frequently to keep your pigs protected. Antibiotics, usually penicillin, are the only treatment.

Swine influenza

Despite recent events involving "swine flu," this is not generally a disease that will infect humans. The media had a tendency to exaggerate the way it transmits to people. Symptoms are much as you would expect with influenza. Pigs will be lethargic, have a fever, cough, sneeze, and show a lack of appetite for their feed. It seldom kills animals, but ill pigs become more susceptible to other diseases, and they will also lose weight while they are sick. Annual vaccinations can keep pigs from getting influenza, but no treatment really exists once they have it.

Atrophic rhinitis

This is a common disease among pigs, and it is usually a mild condition that will not cause problems. It is inflammation of the nasal tissues that can develop into a serious deformity of the nose and snout. Dusty barns with poor ventilation can cause rhinitis, which usually clears up on its own in a couple of weeks. But if there is a secondary infection due to unclean living conditions, the more serious symptoms of a twisted snout can form. Severe cases can be treated with antibiotics, but a clean barn and pasture will usually prevent the disease from becoming serious.

Foot and mouth disease

Foot and mouth, also sometimes called hoof and mouth, disease (FMD) is a serious problem that will require a great deal of diligence if it is endemic to your area. It is contagious, and there is no treatment. Any pigs that develop symptoms will have to be slaughtered, and with any luck, you can prevent it from taking your entire herd. Vaccinations are available but might need to be administered twice a year to be effective. If the disease has outbreaks near your farm, you have to be careful who and what comes onto your property. It is not uncommon to have foot dips and disinfectant for anyone coming or going. Transmission to humans is rare, and it is not usually a serious illness if it is transmitted to humans. Symptoms of FMD include sores on the hooves and mouth/snout area. Before the sores appear, they might stop eating or have difficulty walking. Shoats may die before any symptoms appear.

Butchering

As mentioned in the chapter on butchering basics, most small-scale farmers will hire an outside company to do their slaughtering and butchering to avoid having to meet USDA standards and inspection requirements. However, if you do intend to do this chore on your own, the following are some basic instructions to help you along.

Humane way to kill

The typical steps to humanely kill a pig is to first render it unconscious, and then slice the jugular vein for quick bleeding out. Gassing with carbon dioxide is one way to knock them out, but it is not a practical approach for most small-scale farms. The more common route is to use a captive bolt pistol to the forehead, which immediately stuns them. The animal is then hung upside-down, and the neck is cut with a sharp blade. They will die from blood loss very quickly, and it is all a relatively pain-free and stress-free process for the pigs. Even if you are not going to do this task yourself, you should find out what the local slaughterhouse's policies are so you know that your animals will be treated humanely.

Cuts of meat

Optimal size for slaughtering pigs is between 200 and 250 pounds, and an animal that size will provide you with about 150 pounds of meat. You typically butcher a pig between 6 and 10 months old, but the size is a better gauge. There is a trick to estimating the weight of a pig without having to get it to walk out on scale.

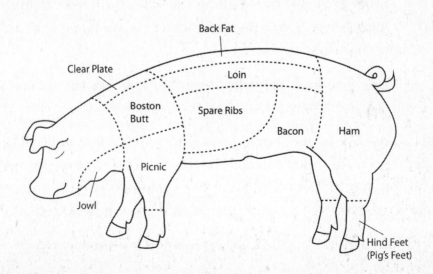

First, measure the girth, which is a measurement around the pig just behind the front legs. Multiply that number by itself (the girth squared). Then measure the length from behind the ears to the base of the tail. Multiply the girth squared by the length, then divide by 400. This will give you the weight of your pig in pounds.

A slaughterhouse will do all the cutting to leave you with all the proper cuts of meat that your customers are familiar with from the grocery store. There are several dozen potential cuts of meat on a pig, and it is not practical to outline each one here. A pig generally is divided up into four major sections, and certain cuts or types of meat will come from each one.

- **Loin:** The loin of the pig is mostly made up of the back area and where you find pork chops, rib roasts, back ribs, tenderloin, and loin roast.

- **Shoulder:** This section is the front area, above the front legs but behind the head. Here are the blade roasts, shoulder roll, and picnic roast.

- **Side:** This is the area along the sides but lower than the loin region. Spareribs and bacon cuts are taken from this section.

- **Leg:** The back haunches or upper leg area has the hams and leg cutlets.

Your slaughterhouse should be able to provide you with all the specifics on the exact cuts and how much of each type of meat you can expect. Depending on your customers' needs, you can also have many of these cuts turned into ground pork instead.

As a rough guide, you will get the following from one pig:

- 46 pork chops
- 4 roasts
- 2 hams
- 16 pounds of bacon
- 6 pounds of spare ribs
- 20 pounds in other cuts or ground meat

Pigs for Profit

You have now reached the point where you can actually sell a product and earn some profit on your hard work.

Selling meat

You can avoid many of the issues relating to specific cuts of meat if you are only selling your pigs by the whole, half, or quarter animal. Taking this approach means you are not selling meat and, therefore, will not require a meat seller's license. Technically, you are selling live pigs. You can then help your customers organize a trip to the slaughterhouse where they can collect their meat.

Although it might be more practical, not all customers are going to want that quantity of meat. After some time in operation, you might be able to build a list of regular customers who do wish to buy this way, but it seldom works out at the beginning.

Your alternative is to obtain the proper license and have the meat processed into cuts for individual packaged sales. Your pigs will still be slaughtered and cut elsewhere, but the meat would be delivered to you and stored for later sales. Everything will be securely wrapped and ready for the freezer. This approach will appeal to more customers, but it will take many more sales to equal the sale of a half or quarter animal. Even with more customers, this way of selling your pork can lead to more waste and lost inventory because not all cuts are as popular. If all your customers are buying hams, you can find yourself with a freezer full of chops that will not remain saleable forever. Pork that has been frozen can be used for up to a year, but customers are not going to be satisfied with buying a product that is older. After about three months, you will have difficulty selling your meat.

Pigs may not be the first animal that comes to mind for small-scale farming, but there are many positive aspects to raising pigs, and they should not be left out of your considerations.

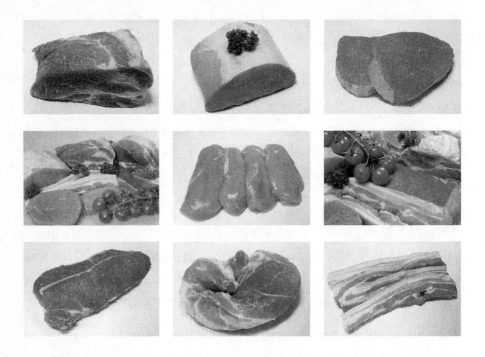

Chapter 6

Chickens

Learn the Lingo

- **Hen:** A female chicken that has begun to lay eggs

- **Rooster:** A male chicken of breeding age, also sometimes called a cockerel

- **Chick:** A general term for young chickens of either sex

- **Pullet:** A young female, before she begins to lay (usually under a year)

- **Broodiness:** The instinct that a hen has to sit on her eggs once laid

- **Vent:** Single body opening for a chicken to excrete waste

- **Combs:** The fleshy growths on top of a rooster's head

Chickens are livestock animals that nearly anyone can raise, whether they are running a true farm or not. They are a great place to start if you are new to keeping animals for food.

Pros and Cons of Raising Chickens

There are several advantages to raising chickens, which is why they are so popular as a "starter animal." One of the main positive aspects of keeping chickens is that, through egg production, you can have an ongoing harvest without having to do any slaughtering. The same can be said of running any sort of dairy operation. However, chickens are even more convenient than milk animals because you do not have to worry about keeping them pregnant in order to produce. Chickens will lay eggs every few days with no effort on your part.

Compared to other animals, chickens are much smaller and will require smaller buildings and less pasture space. If you are running a farm that has limited space, they often will work out well. Chickens are a bit easier to handle due to their size, which can be a plus for someone tending to his or her farming operation alone. On the other hand, they are not the brightest of animals and can be difficult to lead or move around.

They also represent a smaller profit potential because of their size. Chickens are more vulnerable to predators because of their size. A cat or raccoon could not do much harm to a herd of cows, but they could devastate a flock of chickens. You will need to be a little more diligent with security if you keep chickens.

Another positive feature about raising chickens is that they are often not as strictly regulated in terms of inspections and certifications and might be permitted in some areas where other livestock animals are not. *More detail will be provided on this later in the chapter.*

Housing

Unlike most other animals, you will want to keep chickens in a small building to themselves rather than house them in a larger communal barn. Their flighty nature can make the larger ani-

mals a bit nervous, and you cannot have them laying eggs where the eggs might get stepped on. Therefore, you will want to have a separate chicken house or coop for your flock.

The coop does not need to be large; about 4 square feet of floor space for each chicken should suffice. There should also be a foot of perch space per bird as well because they do not like sleeping on the floor and will be much more comfortable if they are able to perch.

Within the coop, you will also need nest boxes. Because eggs will be collected at least once a day, it is not necessary to have one nest box for every hen. They will happily lay their eggs in any free box, whether other hens have used it or not, as long as it is not occupied. One box for every four chickens will usually suffice. A nest box should be about 1 foot cubed and have clean bedding material, such as straw, inside. If you are using a different bedding material, such as wood chips, for the rest of the chicken house, you will still want straw for the nest boxes.

Hens prefer the confines of nest box for laying eggs, so they will naturally congregate there to lay. Not all chickens will conform to this, so check the bedding and even the outside pen for eggs periodically as well.

Of course, the nest boxes are only necessary if you are raising chickens for egg production. Meat chickens are usually slaughtered before they reach egg-laying age, which makes boxes unnecessary.

Because the building will be much smaller than a barn, you need to accommodate for access by both you and your chickens. If you have a coop that is tall enough to walk into, then you simply need a door for yourself and a small lower door for your chickens to get in and out of their outdoor pen. A lower building will mean you have to add doors that will at least allow you to reach in and collect the eggs. It is usually more convenient to build the coop with human access in mind, though.

Your chicken building will need to be well built, with no open cracks or gaps anywhere. Rats and snakes are significant predators for eggs and can get in through even the tiniest opening in your coop. All vents and windows should have metal screening, and you might want to build the house up on blocks to prevent rats from chewing through the floor.

The outside area for your chickens will need more fencing than for other animals, both to keep predators out and to keep skittish chickens in. Most breeds of chickens used for commercial purposes are not able to fly, but some would be able to flap their way over a 4-foot fence if they wanted. Fencing that is 6 feet tall would be a safer option. Consider putting a layer of chicken wire or netting over the top, which will ensure that no chickens can get over the fence and will add protection from flying predators, such as hawks.

Electric fencing is not ideal for chicken pens because you would have to run too many lines to keep them from hopping between the wires. For smaller outside areas, there is electric netting that can be used instead. Otherwise, traditional fencing works fine. Chickens will not usually put too much effort into getting out of a fenced area.

Breeds

Depending on whether you are raising chickens for meat or for eggs, you will want to pay attention to which breed you choose. However, this is not as crucial as you might think. Commercial chickens have been bred more often to be a dual-purpose bird, which means most breeds will be suitable for either production intention. There are hundreds of chicken breeds available, but the following is a list of the more common types found in commercial production.

Breeds that are good at sitting on eggs are important if you want to hatch chicks, but it is actually a negative trait for egg collecting in general. A broody hen can be aggressive while you are collecting, which would make your daily chore difficult. Hens that ignore eggs are better in this regard.

Australorp

These dual-purpose chickens lay frequently and have a heavy body as well. They have black feathers with red combs, and will produce light brown eggs. These quiet birds will not be stressed if you are keeping other animals nearby.

Rhode Island reds

The Rhode Island breed actually comes in either the red or white variety, named for their feather color. The red variety is the more common, though they are equal in terms of production and temperament. They lay brown eggs with a high frequency and are large enough to be used for meat if you prefer.

Leghorn

This is another classic breed that is common around farmyards. This is one that will lay white eggs, which might make it more appealing to some customers who still shy away from brown eggs. Appearance-wise, they come in a variety of feather colors, and they lay frequently. A leghorn hen will seldom sit on her eggs, which makes collecting easy as long as you are not trying to actually hatch out fertilized eggs. They can be a nervous breed and will definitely need a high fence to keep them contained.

Araucana

The Araucana is unusual, and if you are looking to offer eggs that stand out from your competitors, you might want to look into this breed. These chickens lay blue eggs. They are completely edible, and in taste, they are similar to standard white or brown eggs. Araucanas make good novelty chickens, but they do not lay frequently enough to be practical as the sole breed for your farm.

Orpington

The Orpington may not be as well known to the public as the above breeds, but they are great birds for the farm. They are a heavy-bodied bird and will seldom take flight even for a few feet. If you are breeding chickens, you will find that Orpingtons are excellent mothers and will sit on any eggs they can find.

Plymouth Rock

A frequent layer, the Plymouth Rock has the body for a meat chicken as well. They are friendly and docile birds, which make them ideal for first-time chicken keepers. Rocks come in various feather colorations with  the barred (striped) feathers being the most familiar to the point of this breed being sometimes called the Barred Rock. Their eggs are large and a pale shade of beige.

Wyandotte

These mid-sized birds will lay you an egg every other day (or more). They are large enough to be used as a meat bird as well. Unlike a Leghorn, a wyandotte hen most certainly will sit on her eggs while waiting for them to hatch. That is a helpful trait if you want to raise more chicks without an incubator. Friendly wyandottes come in a wide range of colors, and they all lay brown eggs.

Jersey Giant

As the name likely suggests, these are large chickens. Roosters will weigh around 13 pounds and the hens at least 11 pounds. Their size makes them a good meat bird even though they are relatively slow growing compared to most others. They lay large eggs and are docile birds.

Bantam

Although not a breed in its own right, the term "bantam" is used often enough that it warrants explanation. A bantam is just a small variety of chicken that can be found among many different breeds. There are bantam Rhode Islands or wyandottes, for example. Although bantams can be popular among backyard chicken farmers, they are not a good choice for anyone trying to operate a profitable operation. For commercial sale, you are going to get much better results, and profits, from regular-sized birds rather than these miniatures. Typical customers are less likely to be interested in such small birds, as a chicken is not all that large in the first place. Miniature chickens might sell as a novelty, but they will not likely make a solid product for repeat sales.

Equipment

The main pieces of equipment you will need for keeping chickens are the proper containers for food and water. Any shallow container will work for food, and you can get specially designed waterers for chickens that will work better than simple open dishes. They resemble an upside down bottle or bucket in a large shallow pan. The bottle portion is filled with water, and water is dispensed into the pan as it is used.

Chickens will stand in their food dishes if there is room, which leads to a lot of fouled food. Having a trough that is raised off the ground or that has a partial cover will keep the birds out. It will reduce spillage, too, as birds have a habit of picking through grain and flinging it around.

Photo courtesy of Little Creek Farm in Ocala, Florida

For operations that plan to hatch fertilized eggs, either from your own chickens or purchased eggs, you will want to consider purchasing an electric incubator. It is possible to hatch your eggs naturally using your hens, but this can be a risky way of doing it because many commercial breeds have had their mothering instincts bred out of them. Eggs left to hatch in the chicken coop are also at risk from predators or disease.

The safer way of hatching eggs is with an incubator. Smaller units are simple to use and will maintain an even temperature and humidity, but you will have to turn the eggs several times each day so the chicks develop properly. If you are going to be doing a large volume of hatching, a more sophisticated incubator will turn the eggs automatically.

Feeding

Chickens are often thought of as grain eaters, but they are actually omnivorous and will be healthiest with a mixed diet. With a large enough outside area, they will do well for themselves catching worms and insects to add protein to their diet without any additional help from you.

Commercial chicken feed can be purchased in pellet or crumbled form that includes a balanced mix of components to cover all the dietary requirements. You can also buy less processed food in the form of grain and seed mixes that work well for chickens. Do not use birdseed formulas intended for wild bird feeders or pet birds, though. They are too high in fatty seeds for your chickens.

On average, a hen that is laying eggs will need half a cup of commercial chicken feed every day. This is assuming no other food is available. If your chickens have access to pasture, they will need much less additional food.

Fruit and vegetable kitchen scraps are also good treats for the chickens as long as you know they have come from organic sources, such as your own garden. Chickens are also known to enjoy cooked pasta and stale bread, but those can be tricky to manage unless you have organic food to offer.

Aside from actual food, you will need to provide two other things for your chickens to keep them to healthy. The first is grit.

Because chickens have no teeth and still eat hard foods, such as dried grains, they need a way to crush up their food. They will instinctively eat small rocks or pebbles and store them in their gizzard to aid in digestion. The stones eventually wear away and will need to be replaced, so it is an ongoing process. Depending on the size and nature of their outdoor space, you may not need to provide this for them, as they will be able to find small stones on their own. However, after many months of chickens picking up pebbles, they might actually run out if they have a small outdoor area.

The next item you might need to supply for your chickens is additional calcium, though only for hens that are laying eggs. Feeding a commercial pellet feed might make this unnecessary, as most formulas will have extra calcium for laying hens, but if you are feeding your own mixtures of grain or letting your chickens free-range for their food in a pasture, you will need to give them extra calcium. Crushed oyster shells are a common supplement for this. A traditional approach is to feed eggshells back to your chickens, but your hens might be more likely to peck and eat their own eggs if you do this.

Breeding

You will want to breed your hens if you want eggs to hatch into chicks. You do not *have* to keep your hens pregnant or bred in order to have eggs. This is quite a misconception when it comes to chickens, and the topic should be understood before considering a breeding program.

Breeding basics

One great thing about chickens is that you do not have to time your breeding plans to coincide with a female's reproductive cycle. A hen's eggs can be fertilized at any time, so that will simplify things for you. Sperm from a rooster can be held inside the hen up to a week to fertilize several eggs. A hen will create eggs every one to three days no matter what, and if there is active sperm available, the eggs will be fertile. Only fertile eggs can hatch into chicks, and usually unfertile eggs are used for commercial sale.

Many agricultural companies sell fertilized eggs or even day-old chicks if you prefer to avoid the entire issue. If you have a large operation, this might actually be the better course of action when you need to add new birds to the flock.

How you organize your breeding program will be somewhat similar whether you are keeping chickens for eggs or for meat. Meat chickens usually are slaughtered before they reach maturity or breeding age, so you will need to have a separate population of hens and roosters old enough for breeding. This is also a convenient method when dealing with egg-laying chickens, as it means you do not have to distinguish between fertilized and unfertilized eggs.

Meat operations will have to maintain a higher number of fertilized eggs to make up for the loss of chickens due to slaughter. On the other hand, egg operations will need far fewer new chickens, as hens can continue to lay for three or four years before you have to replace your hens with fresh birds.

Egg incubation

The simplest way to incubate your eggs is to just leave them out with your hens and let them take care of the entire job. This only will be an option if you are raising chickens that still have some of their natural mother skills intact. Because collecting eggs from under a broody hen can be difficult, the instincts have been bred out of many commercial chicken breeds. For a properly attentive mother, you will have to have breeds such as Orpingtons, wyandottes, or silkies. You can also maintain a population of these breeds to act as surrogate mothers if you prefer to raise other breeds for their eggs. They will happily sit on any eggs, whether they laid them or not.

A more precise method for incubating eggs is with an electric incubator. It eliminates any risk from outside predators or inattention from your hens. Incubators will keep your eggs at the right temperature as well as the right humidity, so read the instructions on how to add water. There is more to running an incubator than plugging it in and setting the temperature. Most incubators will need to be set at 99.5 degrees Fahrenheit, but this will depend on the model and how the fans move the air. Your instruction manual will tell you the proper levels.

The incubation period for chicken eggs is 21 days. For the first 18 days, you will need to turn your eggs twice a day if your incubator does not do it automatically. A little pencil mark on the shell will help greatly when keeping track of which eggs have been turned and which have not. Normally, this is something the hen would do several times a day in the nest. It keeps the developing chick from settling to the bottom of the eggs and sticking to the shell. For the last two or three days, you will need to stop turning the eggs so the chick can orient itself to know which way is up for safe hatching.

Not all of your eggs will be fertile, so check your eggs after three or four days in the incubator. The process is called candling, though you no longer need to use a candle to do it. All you need is a bright light so you can see a silhouette of the inside of the egg. You can buy small lighting devices for this if you wish, or you can make do with a small flashlight in a darkened room. Place the light against the shell and hold the egg about a foot away from you at eye level. This should illuminate the inside of the egg brightly.

It can take some practice to know what you are looking for and some photos or an experienced eye to help. You will want to see a small dark mass in the middle, with a pattern of veins around it. That is the developing embryo.

Eggs that are clear after five days are probably infertile and should be taken out of the incubator. You will also want to remove any eggs that are showing a faint ring inside. This usually means bacteria has gotten into the shell and is working its way through the inside. It will go bad quickly and can possibly contaminate other eggs.

Once the eggs begin to hatch, you will have to let the chicks do the work. With a little experience, you might be able to assist if necessary, but a novice should not attempt to "help" a chick out of their shell. You will usually do more damage in the process.

The downside of incubating eggs is you will also have to play mother to the young chicks until they can be released into the chicken yard. You might be able to move eggs from one chicken to another, but hens are unlikely to care for chicks they did not hatch themselves.

Health

In a flock of chickens, it can be difficult to watch and monitor each individual bird for health problems. However, the sooner a health matter is noticed, the sooner you can treat it, so make an effort to get to know your birds and give them some type of inspection regularly.

Newborn health

Newly hatched chicks will need a carefully controlled environment, as they normally spend their first days and weeks under a hen. You will need to take care of them for a couple of months before they can be released out on their own. Once you have created an area to keep your chicks safe and warm while they are growing, the time you will have to put into their care is minimal.

After they have hatched, leave your chicks in the incubator for one to two days. They will not need any food or water during that time, and it gives them a chance to rest before being moved.

The enclosure used for young chicks is usually called a brooder, and it is just a secure area where they will be kept warm. A large box works fine, as long as the sides are high enough to keep the little birds from hopping out. Keep them warm with a mounted heat lamp with an ambient temperature about 90 degrees Fahr-

enheit. After the first week, you can move the lamp farther away and drop the temperature by 5 degrees each week. The brooder should be close to room temperature by the time your chicks have their feathers, which is when you put them out with the other chickens. It is usually a good idea to move them out slowly, over the course of a week or two; so your existing flock can adjust to the newcomers.

While your chicks are in the brooder, feed them starter feed, which is designed for chicks and comes in a powdered or finely crumbled format. Most chicken starter comes medicated, so make sure you have a brand suitable for organic chickens. Use a shallow saucer for water or chicks can easily drown.

Other than basic chick care, you must look out for other health issues as well. One common disease is the pasted vent. In their first few days, chicks with this disease will have runny droppings. With very fluffy feathers, this can be a problem, even in clean surroundings. Examine your chicks regularly to make sure their vents are not clogged with wet feathers and waste. This can kill a chick within a day or two, so do not treat it lightly.

One final health concern that specifically affects chicks is spraddle leg, which is more a result of brooder conditions than an actual disease. If young chickens are kept on bedding or flooring that is too smooth, they will have difficulty standing up because their legs slide out from other them. As the bones in their legs harden, they can be permanently deformed. Always keep chicks on a rough surface, such as wood chips. Newspaper can be a problem once it gets wet.

Chicken diseases

Because of the relatively quick turnover time with raising chickens and the low investment in each bird, many farmers do not spend a lot of time, money, or effort treating chicken diseases. Birds are often killed when they get ill rather than treated. However, this does not mean that vaccinations and other treatments are not available for chicken farmers.

Vaccinating your young chicks will help protect against future losses, and you can treat many diseases if you watch for the symptoms early enough.

Salmonella

Salmonella is a common buzzword when it comes to diseases in egg facilities, but it usually affects your eggs — and your customers — rather than your actual chickens. As long as you keep your chicken enclosures clean and practice good hygiene when collecting and storing your eggs, you should not have serious problems with salmonella. The bacteria are common and can cause your chickens to fall ill if they are too numerous or your chickens are otherwise unhealthy. A chicken with salmonella will have symptoms including diarrhea and lethargy, and any ill chickens will be off their feed. Normally, treatment would include antibiotics, but it is seldom fatal. You can isolate the sick chickens, give the main enclosure a thorough cleaning, and just let them get over it if you do not want to risk their organic standing with antibiotics.

One particular variety of salmonella can cause pullorum disease in young chickens. Thankfully, it is not that common, as it has been eradicated from most commercial flocks. White diarrhea, ruffled feathers, listless behavior, and limping are all symptoms in chickens under a month old. Older chickens seldom get ill. Antibiotics are the only treatment, which cannot be used on your organic farm, and vaccinations are not used because it can encourage the creation of "carrier" birds that further spread the disease.

Fowl pox

The pox that can be transmitted between chickens is not the same as the well-known chicken pox disease that affects people. Insects spread it; so it will likely only find its way into your flock if other birds in the area have it. There is little you can do to protect your chickens from biting insects in general.

Black sores will form on a chicken's combs, but usually, the disease will not create other severe symptoms beyond that. Hens with pox will stop laying eggs until they are no longer ill. Although it is not a fatal disease, it can disrupt your egg production and can be vaccinated against.

Coccidiosis

This is usually only a threat to younger chickens, so you need to keep an eye on your chicks and younger birds. After this period, they usually have enough natural immunity to fight off any infections. Listless birds with bloody diarrhea might be the first symptoms, but by the time you see symptoms, it might be too late to do much for the birds. The bacteria live naturally inside chickens, which means it is always a potential problem with confined flocks of birds. Keep the enclosures clean and sterilize the water containers frequently to help keep the bacteria population down.

Avian flu

There is a lot of misinformation about avian flu in the media, so take care to read about it before making any rash judgments. Symptoms for avian flu are much like the symptoms found in humans with similar illnesses, including lethargy, coughing, wheezing, and lack of appetite. It is generally not a serious disease, but you can vaccinate your chickens if it is a problem in your area. It is not a disease that commonly transfers from chickens to people, regardless of what you might have read in the newspapers. Eggs can be collected from ill chickens as long as they are washed well. You will find that ill hens will lay fewer eggs until they have recovered.

Marek's disease

Not all chicken diseases are as harmless. Marek's disease is quite serious and usually fatal. Vaccinations are highly recommended. It is a nervous system disease that will cause paralysis and death. One visible symptom is odd coloration in the eyes. It spreads through dust, so it can be transferred from one farm to the next, but it will not spread great distances as an insect-borne disease might.

Impacted crop

This is not a disease but a condition that can affect chickens and can be fatal if not taken care of. As mentioned in the feeding section, chickens store small stones in their crop or gizzard in order to help break down the food they eat. This crop is like a small sack in their throat, and food such as hard grains or straw are temporarily stored here while being digested. If a chicken is eating too much food that is hard to digest (straw in particular) it can clog up the crop. It will get hard and too full to allow any new food to be passed through. Your chickens will stop eating and eventually will die if the crop is not cleared out.

A little mineral oil forced down the throat can help quite a bit, and a vet can be called if some minor surgery is needed. It can actually be done yourself once you have watched the vet do it. A small incision to the outside of the neck will open up the crop and let you pull out the compacted material. If this happens with your birds, re-evaluate the bedding and food you are using.

Butchering

Given the small size of a chicken, it is not practical to butcher them into smaller cuts of meat for resale. Other than slaughtering and cleaning them, which involves removing the feathers, skin, and organs; there is little butchering to do with chickens unless you decide you want to sell individual cuts of meat from your birds.

You can slaughter your meat chickens after about eight weeks, but the time for slaughtering depends on how large a chicken you want to sell. At 3 pounds, they are referred to as "broilers," whereas a "roaster" is closer to 5 pounds. The sizes and terminology are not set in stone, though, so you can use your own judgment on size depending on what your customers want. A chicken that is about 5 ½ pounds will provide a cleaned bird of 4 pounds.

Slaughtering chickens is a chore that most small-scale farmers can do themselves, but you will find the legalities easier to deal with if it is done by a professional, as is recommended for other livestock as well.

Egg Collection

For chicken operations that produce eggs rather than meat, it is the egg collection that gives you a harvest rather than slaughtering. This means more ongoing work to manage your flock.

High-frequency layers can produce one egg every day, but that level of production is not typical. Good layers will give you an egg every two or three days on average. This means you are going to need to collect your eggs at least daily in order to get them while they are still fresh. During the summer, it would be prudent to collect eggs twice a day so none are sitting in the heat for too long.

Breeds with strong mothering instincts, such as the Orpington, are more likely to sit on their egg once they have laid them. This can make it a little easier to know where the eggs are, but you have to contend with the hen sitting on them. She will not be happy when you steal her eggs. Other breeds will just lay and leave, so you will need to check all the nest boxes to see if eggs are present.

One aspect of egg collection that has yet to be mentioned is the reduction in laying that naturally happens during the winter. The shorter days trigger a slowdown in egg production, and some chickens will stop laying completely. If you are not selling as many eggs during this time, you can let your flock have a rest until spring. Otherwise, you will need to lengthen the day artificially to counteract this natural process.

Chickens need 14 to 16 hours of light each day to maintain their peak laying cycles. Rather than wait until the days are getting shorter, have a lighting setup already in place. Even just a few short days can trigger a slowdown, and it can take several weeks of light to bring production back up. All you need is a light or two in the enclosure that is activated by a timer. The best approach is to have the light come on earlier in the day rather than stay on later at night. The sudden shutoff of the light at night can be disturbing for chickens caught unawares in the dark without a roost. Because they need to be perched on something to sleep comfortably, they need a chance to get settled before being left in the complete darkness.

Once collected, you will need to clean your eggs. This is actually not quite as simple as it sounds. Eggs are naturally covered with an antibacterial coating called the bloom. Scrubbing your eggs too hard will remove this and can even force germs through the eggshell and make matters worse. First, try to wipe them clean with a dry rag to remove any obvious dirt. If the eggs have manure on them, you will have to use soap and water to clean off the shells.

Chickens for Profit

Chickens might not earn you quite as much per chicken as the other large livestock choices, but you can definitely have a profitable operation selling either meat or eggs.

Selling meat

You will only need to get a state license to sell chicken meat if your farm is producing more than 1,000 chickens per year, though you could also bypass the need for a license in the same way as you do with other animals — by technically selling the entire live bird to be processed by the customer. It is not as common an approach with chickens due to their small size.

There might be more sales options if you have the chickens butchered into smaller cuts, but there is little difficulty in selling whole chickens either. Once you get to know your customers a little better, you might want to offer some smaller cuts, such as the breast, legs, thighs or wings. However, you will risk having wasted meat if you only sell certain cuts and have the others left on hand.

Selling eggs

Selling chicken eggs is the least regulated way to make money from your livestock, but the exact regulations will vary by state. In some states, as long as you are selling eggs directly to your customers, not selling them to other retailers, you will not need a license at all. In other cases, the size of your operation will determine whether you need a license; you might be exempt up to 200 or more hens. If you do require a license, the process is usually simple and inexpensive.

You can apply to have your eggs graded by the USDA, but this will add considerable expense. As long as you clearly mark your eggs "ungraded," inspections are not required.

As long as the eggs are clean, all you will need to do is package them. Many regular customers will be happy to bring back their empty egg cartons, or you might need to purchase cartons for your eggs.

Chickens are a classic farm animal that take up so little room that nearly any operation can afford to have a small flock of laying hens to provide a few extra eggs. Even if you are raising other animals, chickens will usually fit in somewhere.

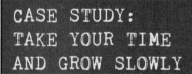

CASE STUDY:
TAKE YOUR TIME
AND GROW SLOWLY

Julie Rawson
Many Hands Organic Farm
411 Sheldon Road
Barre, MA 01005
www.mhof.net

After growing up on a conventional farm, Julie Rawson continued the family tradition but chose an organic path instead. Her personal life-long struggle with chemical sensitivities made it a natural decision once she discovered the idea of organic farming back in 1970.

The name "Many Hands" is an apt one. Their farm has taken several different approaches to raising livestock as well as flowers, fruit, and vegetables. In the beginning, they started out with just a few turkeys. That was back in 1984. Today, the animal population at Many Hands is large and diverse. Several hundred chickens and turkeys, as well as a handful of pigs, call the farm home. The birds are raised for meat, but some birds are dedicated to egg production as well.

Most of their sales are done through various CSA programs, where customers buy a share of her overall production for a season. They also use several different farmers markets around their Barre, Massachusetts, location.

Rawson said they are happy with where the farm is at right now, but she does have some plans for the future. She said, "We just want to become more efficient and raise more of our feedstuffs on the farm, sprout more, use whey from our certified organic neighbor, and improve our pasture."

She has found the overall process of becoming certified easily managed, highly intuitive, and without any serious roadblocks along the way. Finding organic supplies and feed have not been a problem, and she manages to care for her entire farm without the use of a vet. Many Hands does use an outside slaughterhouse for the butchering chores, though.

For anyone looking to begin an organic farm, Rawson offers a little personal wisdom, "One venture at a time; do not expand too quickly. Understand what you are doing before you move forward. Sell to the end buyer to keep the profit level as high as possible. Buy only good stock. Make animal health through proper pasture, high-quality feed, diverse feed sources, proper stocking rates, frequent pasture moving your highest priority."

Is there anything Rawson would do differently if she had the chance? Not really. She has learned as she went along and has enjoyed the process because wisdom only comes from age and experience.

Chapter 7

Horses

Learn the Lingo

- **Stallion:** Male horse of breeding age

- **Mare:** Female horse of breeding age

- **Broodmare:** A female whose sole purpose is for breeding, rather than racing or working

- **Filly:** A younger female horse

- **Colt:** A younger male horse

- **Foal:** A young horse of either gender that is still with its mother

- **Gelding:** Male horse that has been castrated

- **Foaling:** Giving birth

- **Farrier:** A professional who shoes horses and cares for hooves

Horses are the only livestock animals included in this book that are not raised for food production. Nevertheless, their use and value as sporting, working, or recreational animals make them potential money earners.

Animals raised for food usually have a lifespan of under a year or two, whereas a horse can live longer than 20 years. There is a far greater investment involved with a horse, and the level of care is going to vary in accordance with that.

This book has focused on the requirements for raising organic animals, but the concept does not apply in the same way for animals that are not raised for food. Whether you choose to have your farm certified or not, you can always follow organic principles when it comes to the feed and medication you use with your horses. Many certified organic food products are available, as well as other products you can use for your horses, such as medications.

Pros and Cons of Raising Horses

One significant downside of raising horses is the cost. They are generally much more expensive to buy and maintain, especially if you are going to keep them for several years before selling them for a profit. Training will add to your expenses if you intend to sell your horses for specific purposes. Conversely, the profit potential also can be higher.

They are also much less hardy when it comes to grazing. In other words, a horse will not eat the same variety of plants a goat or sheep would. This means higher feed costs for you or higher maintenance expenses to keep pastures properly seeded.

That is not to say there are no advantages to breeding horses. They are generally a little easier to handle and can be trained better than other livestock. Horses are intelligent, and they have been bred for their training abilities more so than livestock animals bred for food production. Because you will not raise these animals for food, you will also be free of all the red tape that such activities usually involve.

Housing and Pasture

Large animals such as horses will need much more space inside their enclosures not only because they are larger but also because they do not tolerate crowding the way some smaller animals do.

Stalls for horses should be at least 12 feet by 10 feet in size, what most prefabricated stall units are constructed as. It is typical to allow one stall per horse, though mares can be housed together with their foals. A horse barn should protect your animals from any drafts but also be well ventilated so moisture does not build up inside the building. Bedding material can be straw, sawdust, or wood chips, and it will need to be cleaned out daily.

Fencing for horses is usually electric or contains at least one wire of electric fencing to prevent the animals from pushing on a wooden or woven-wire fence. Because horses are more likely to

be on the run than other animals, it is important to make any wire fencing, especially a single line of hot wire, visible so the horse can see it even when moving quickly. Using wide electric ribbon instead of wire is a preferable approach to prevent entanglements or injuries.

Horse pasture is just as important for exercise as it is as a source of food, so you need to allow more space per animal than you would otherwise. On average, 1 acre of outdoor space will be sufficient, but you might need closer to 3 acres if you plan on using your pasture as a major food source for your horses. If you are allowing your horses to get exercise away from their main paddock, such as on riding trails or a jumping ring, you can get away with a smaller amount of space at home.

Stallions cannot be kept together because they are aggressive and will always fight, so you will need to allow for a separate paddock for your male horses. Geldings are fine to be housed together or with your mares.

Breeds

Because there are several different uses for a horse, it is not surprising that there are many different recognized breeds that reflect these different characteristics — several hundred breeds, in fact. Stamina, speed, size, color, and temperament are all variations that are important from one breed to the next. Generally, you can divide the breeds into two groups: sporting breeds and working breeds.

Sporting breeds

The term "sporting" in this instance refers to horses that are used for either racing or agility competitions, such as barrel racing or jumping.

Thoroughbred

The term "thoroughbred" has been used to describe any pure-bred horse, but it is its own specific breed known for its excellence in racing. Thoroughbreds are also used in show jumping, dressage, and polo. They have been frequently bred with other breeds to improve their bloodlines. The quarter horse and many warmblood horses began as crossbreeds with the thoroughbred.

These can be valuable horses, but specific pedigrees and performance play a large role in determining the value of any one individual horse. The most expensive horse sold at auction was a thoroughbred named Green Monkey, which sold for $16 million dollars. Famous racehorses Man o' War and Secretariat were both thoroughbreds as well.

As a breed, they are known to be a riskier investment due to prevalent health problems. Low fertility and bone fractures are common problems, and thoroughbreds can have bleeding in the lungs in some cases. This is one reason why successful racehorses are so valuable.

Arabian

 Descended from the desert horses of the Bedouin, the Arabian is another well-known breed in the racing world, particularly in longer races that make use of their natural endurance.

As with the thoroughbred, the Arabian bloodline can be found in many other modern breeds, such as the quarter horse. They are even-tempered and usually easy horses to work with. A few known genetic conditions can affect this breed, but they are not generally prone to health problems as the thoroughbred can be.

Quarter horse

The full name for this breed is the American quarter horse, and it is suited for short-distance racing or sprinting. The breed originated in the United States and is used for disciplines that require short-term speed, such as rodeo competitions and ranch work. Quarter horses are favored as a working horse on ranches where other animals need to be herded or controlled. Their primary use is for racing.

American Standardbred

Originally a mixed breed between the thoroughbred and the Morgan, this is one of the premiere racing breeds being raised today. Their natural gait makes them well suited for harness racing, and they are  popular horses in Kentucky.

Warmbloods

The warmblood is really a general type of horse rather than a true breed. The term, however, is used so often in sporting circles that it bears explanation. These breeds fall between the racing (hot blooded) and working horses (cold blooded). Rather than racing, these competition horses are bred for events such as dressage, show jumping, and other such equestrian sports. The breeds in this group are usually just named for their countries of origin, such as the Dutch warmblood or the Swiss warmblood.

Working breeds

With the advent of motorized agricultural equipment, horses are not used for work as they once were. Nonetheless, the breeds remain and are still in demand across the country. As the interest in simpler farming increases, more people are looking for good working horses for their small farms. Another general name for working breed horses is "draft horse," though this does not refer to any one breed.

Clydesdale

The large Clydesdale is a distinctive breed that can weigh twice as much as a race-horse. The long hairs on their hocks give their hooves a feath-ered appearance, and this is the feature

most people recognize. Their well-muscled bodies are best suited for hauling, and they have a long history in agriculture.

Appaloosa

The Appaloosa is a different kind of working horse that is geared toward cattle herding, roping, and other similar types of chores. Their spotted coat is one of their most distinctive features, though various colorations are possible within the breed.

Percheron

Although they can occasion-ally do well in the jumping ring, Percherons are generally a draft horse used for hauling. They are not as large as Clydesdales, but

they do have sturdier bodies than breeds intended for racing.

Equipment

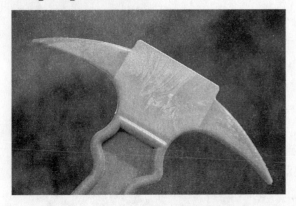

Grooming tools are important to keep a horse's coat, mane, and tail clean and looking presentable. A various selection of combs and brushes are available for grooming, and you can take your pick as to which ones suit your needs. Horse shampoos and conditioners also are available to keep their coats in top shape.

Caring for a horse's hooves will require special equipment. Trimming and cleaning of hooves, with or without horseshoes, is a daily chore that will require a few pieces of equipment in order to do the job right. A professional farrier usually does the actual shoeing of horses, so that job does not need to be your immediate responsibility.

A blunt hoof pick to pry small stones or clumps of mud out of a hoof is extremely important, as are some small files and pliers. The tools generally are used to remove any debris or smooth any roughness that develops on each hoof. Poorly maintained feet can lead to major problems or even permanent lameness from infection. Another product you should have on hand is hoof cream or ointment to keep them pliable and prevent cracks from dryness.

Other important horse equipment is collectively called the tack. This would include saddles, bridles, harnesses, bits, and saddle pads. Your needs for these items will depend on how you are raising your horses for sale. Horses that are sold young, as year-lings, will be mostly un-trained, leaving you with minimal equipment require-ments. However, if you are keeping your horses longer to be trained for a certain pur-pose before sale, you will need the right tools for that training. Even if you are not  breaking your horses to saddle, a harness and lead will be essen-tial for basic handling.

For leather items such as bridles and saddles, you will also want to have the right kind of leather soap and oils to keep them supple and smooth. Fly masks are a handy thing to have if your horses are bothered by flies around their heads during the summer.

You will also need the usual equipment for feeding and watering your animals, usually just large open troughs for water and hay. Racks also work great for hay feeding but are not necessary.

If you are going to ride your horses, a proper riding helmet and boots are common pieces of equipment for yourself. Proper rid-ing pants or even leather chaps can be helpful if you are going to be spending a lot of time on horseback.

Transporting horses often comes into play, whether it is for a po-tential customer or to take your horses to shows and competitions. A horse trailer and a vehicle strong enough to pull it can be important.

Feeding

The bulk of your horse's food will either be dried hay or forage from a large pasture. Added grains or commercial horse feed is another component, though these will only be an addition rather than a mainstay.

Horses have a more delicate digestive system than most other livestock animals and cannot be fed on a simple field of unknown wild plants. In order to maintain a usable pasture, you will need to seed it each year with the proper vegetation. It does not need to be anything fancy. Horses will do well on most grasses, as well as alfalfa. Alfalfa alone can be a little too rich in protein, but you can mix the seed in with the grass. Because you must maintain the pasture in this way, it is a good idea to have at least two fields for your horses so they can graze in one while the other is growing.

If they are not kept on pastures large enough to sustain them, you will also have to feed them hay. You can specifically buy alfalfa hay or even compressed alfalfa pellets as horse feed. Horses favor grains such as oats and barley as additional food sources. Most horses will need about a bale of hay per day if that is their main source of food.

Aside from their pasture grazing, they should be fed several times a day in small amounts. Horses will drink a great deal of water, so it should always be available to them. An average horse can drink up to 12 gallons per day and more in hot weather.

A field full of lush green grass might seem ideal for horses, but health problems can arise if you turn your horses out into such a field when previously they have eaten only dried grasses or hay. This can happen when you move from winter enclosures to outside pastures in the spring, for example. Too much green grass will cause a condition called "founder," or more technically, "laminitis." Oddly, it is not a problem with the digestive system, but rather the horses' feet. The inside of the hoof becomes inflamed, causing pain and tissue damage.

Initial symptoms of founder include a fever and an odd gait as the horse attempts to keep their weight off sore hooves. Horses might breathe heavily, and their feet can be hot to the touch. If left untreated, a horse can have permanent walking difficulties.

Do not allow your horses to free-feed for too long when the pasture is too lush with green grass or when there is water on the grass from recent rain or heavy dew. Once a horse has foundered, it is much more likely to have future recurrences and will have to be restricted to a diet of only dried hay and grain for the rest of its life.

Horses represent a longer time investment, so healthy feeding is even more important. Vitamin supplements are a good idea for the overall health of your horses, as long as you ensure you are using natural products that conform to the organic regulations. Discuss their needs with your vet to see what specific minerals or vitamins your horses need. It will vary depending on the diet you are using and the amount of exercise your horses get regularly. Hard-working horses will have different needs than those that spend their days quietly grazing.

Breeding

The breeding aspect of keeping horses is more complex and more important than with other animals because a horse's value is closely tied to its lineage. This is vitally important in the world of racehorses.

Breeding basics

A filly will become sexually mature at about 18 months old, but it is not a good idea to have her bred that young. It is a more common practice to wait until she is around 3 years old before she has her first foal. The gestation period for horses is 11 months (340 days).

A mare's reproductive cycle is about 22 days long, with a heat period between four to eight days when she is receptive to getting pregnant. The remaining days of the cycle, she cannot be impregnated. She will cycle through this during the spring, summer, and fall months but will stop going into heat during the winter due to the influence of the shorter days. When a mare goes into heat, her behavior might change, but this depends on the horse to some degree. They can get moody, irritable, and be difficult to handle. Another common symptom is to urinate around a stallion, which might not be helpful if you are not keeping stallions on your farm.

As with any livestock animals, the need for keeping breeding males on the premises is always an issue. Stallions are too large to house and keep for once-a-year breeding purposes unless you have a particularly valuable champion on your hands. Most horse-breeding operations, especially those on the smaller-scale, use artificial insemination or rent a stallion when the time comes for his services.

If you are breeding thoroughbred racehorses, all registered horses will have the same official birthday of January 1 of the year they were born. This is done to create some uniformity when establishing age classes for races. Therefore, racehorse breeders try to have their foals born soon after the first of the year. Any foals born later in the year will be at a disadvantage. For example, a foal born on January 2 will truly be a full year old on its next birthday when it is considered 1 year old on paper. On the other hand, a foal born on April 1 will only be 9 months old when it is considered 1 year old on paper. A more extreme example would be a foal born in November. They will be considered a year old in January for racing purposes when they are only 2 months old. Therefore, when breeding thoroughbreds, you will want to get your mares pregnant in March or April for a birth close to January.

Prenatal care

Mares that are usually ridden can keep up with their regular routines well into their pregnancies without any problems. As long as you are not working them excessively, you can keep riding her until she is about a month away from giving birth. Just do not forget to adjust the saddle straps as she gets bigger. If you are jumping your mare, you will need to stop jumping four months into her pregnancy.

If you are not riding or working her, she still needs exercise and should be turned out into the paddock or pasture regularly.

Consult with your veterinarian about using any additional vitamins during pregnancy, specifically prenatal formulas. Your veterinarian will need to know about your organic goals before suggesting specific products. Also, vaccinations may be in order so your mare's milk can transfer some of that protection on to the young foal while they are suckling.

Birthing

The average gestation period is 340 days, but unfortunately can vary between 315 days to a full year. This can lead to some stressful times for an expectant horse owner as you wait.

Her udders might begin to fill out as she gets closer to delivery, but if this is her first pregnancy, this might not be noticeable. Other signs are increased urination, pacing, tail twitching, and general signals of discomfort. Lead her into a large clean stall to give birth. She might lie down, or she might choose to stand during the delivery. Do not try to force her to change positions, and be ready to ease the foal down to the ground if she is upright.

Once the contractions begin, she might deliver the foal in as little as ten minutes. After 30 minutes, you should contact the veterinarian if you have not already called him or her to attend the birth. Horses are large animals, and problems during foaling will be difficult for you to handle; so it is highly recommended that you have your veterinarian on hand for the birth if possible or at least a phone call away.

After the foal has been born, it should be up on its feet within an hour and feeding by two or three hours after birth. *Further care of newborns is covered in the next section.*

Health

As mentioned at the beginning of this chapter, care and treatment of a horse will usually be more in-depth than that of other animals due to their lifespan. A closer look at long-term health is as important as caring for immediate needs.

Newborn health

As long as there are no complications, there is little need for you to do anything with a newborn foal. A gentle wipe with a dry towel will help clean it off, and the umbilical cord should come off without any assistance.

Keep mother and baby together in a stall for a day, and make sure it is warm enough for them. Extra heating is a good idea if you are birthing in the winter. After that time, you should be able to turn them out into a small empty pasture for some exercise. It is best to keep them separate from other horses for about a week. After that, you should be able to keep them with the rest of your herd if you have additional horses on the premises.

If your mare does not produce enough milk or rejects the foal, you will have to take over its care. This is not a common occurrence but should be considered a possibility. You will need milk replacer formulated for horses. Bottles can be used at first but even young foals can learn to drink from a shallow bowl to make things easier. Feedings should be given every two hours for the first few days, and you can slowly spread them out after a week to every six hours.

Whether the mare is feeding or you are, you will be able to introduce grains or hay to your foal after about three weeks. Foals kept with their mothers will usually experiment with grass, hay, or grain as they see their mothers doing so.

As long as your foal is still drinking the mare's milk, they should be housed together in the same stall. However, after three of four months, you will want to work on weaning them apart. During these first few months, you also want to spend time with the young horse to get it used to human interaction. In particular, experiment with lifting its feet so that it is accustomed to such treatment. It is important to get the horse accustomed to this so it will stay still during hoof cleaning and farrier treatment.

Castrating male horses, or gelding, is not a procedure done to a young animal, as it is with many other species. Horses do not become sexually mature until about 2 years old, so gelding can take place anytime during those two years without any problems. It is typical to castrate any male horses not to be used in breeding by 1 year of age. It is a surgical procedure a veterinarian should perform. Males that have been gelded will have a much more docile and nonaggressive personality than intact stallions, which makes them suitable for riding or working.

Horse diseases

Depending on how you are selling your horses, they might be in your care for many years before moving on to potential customers. Therefore, you will need to know more about potential illnesses that can affect both young as well as older horses.

Founder (laminitis)

This was covered in the feeding section, but it warrants a mention here as well because there are other causes of laminitis aside from too much green grass. Other infections can cause it, such as salmonella or Cushing's disease, which is mentioned below, and mares are more susceptible to it after they have given birth. Look for symptoms of sore feet and possible fever. Restrict your horse's diet to hay and keep them on a soft surface, such as wood chips. Contact your veterinarian immediately. Prolonged infection can lead to permanent lameness and hoof deformity. Technically, laminitis is the name for the inflammation of the tissues under the hoof, and founder is the resulting lameness or tissue damage. Many people do use the words interchangeably.

Equine influenza

Most animals are prone to some sort of influenza virus, and the horse is no exception. This nonfatal illness will keep horses out of shows or competitions while they are sick, so there can be an economic price to pay. Vaccinations are a good idea to keep your horses from picking it up, as it is a highly transferable disease if your horses mingle with others. Symptoms are typical for the flu: fever, coughing, runny nose, and general irritability. Most horses will be over it within three weeks.

Colic

Colic is extremely common in horses and is more of a symptom than a specific illness. Colic refers to any sort of stomach pain or intestinal discomfort that can be caused by more than one problem. Unfortunately, you will need to know how to tell the various causes of colic apart so you can determine when a veterinarian should be called. It can be caused by a simple case of indigestion, or it could be more serious, such as a case of kidney stones, parasite infection, or twisted intestines.

Watch for a horse that bites at its flanks or belly, rolls around, paces, sweats excessively, and stands in an odd position. First, try to get the horse to walk around slowly to help work out any stomach problems. After about 30 minutes of gentle exercise, symptoms of minor colic should have abated. If not, call the veterinarian.

Tetanus

Tetanus is a common infection that can affect many animals, including humans. Unfortunately, horses tend to be much more susceptible to contracting it. It is usually fatal if left untreated and typically starts with a wound of some sort. Rusty metal is common cause of tetanus. The bacteria will eventually cause paralysis, and the symptoms are related to that developing paralysis. Horses with tetanus will have a stiff gait and difficulty moving, and they might clench their jaws repeatedly. Eventually, the disease will lead to seizures and death.

Treatment includes penicillin as well as tetanus antitoxin, which would nullify any organic standing for the horse. Because it is such a serious illness, vaccinations are highly recommended. You should also regularly check the stables and pastures for any metal debris.

Cushing's disease

Cushing's is not a disease caused by any virus or bacteria but results from a pituitary tumor in the brain. The result is a hormonal imbalance that causes a long, wavy coat of hair along with excessive thirst. You do not usually see this in younger horses, but rather those over the age of 8. There is simple medical treatment for this condition, and it does not often have any impact on your horse's day-to-day life or abilities.

Thrush

Clean stalls can help prevent thrush, which is a bacterial infection of the hoof. When checking the horses' hooves, if it has thrush, you will notice a black discharge around the soft parts inside the hoof and an unpleasant odor. If you are cleaning out your horses' feet regularly, you should catch it right away. With regular cleaning, it probably will not even be a problem. A few applications of a topical antifungal product should clear it up quickly, and this is allowed for organic animals.

Training

Horses are the only livestock option where you may need to have training as part of their upbringing. It is not an issue for animals being raised for food. How much training will depend on when and why you are selling your horses. Young colts are not expected to have the level of training of an older horse, for example.

The most fundamental level of training would be to have your horse accept wearing a halter and walk on a lead. You will want to get your young horses adapted to being handled this way to make it easier to move them about, particularly as you move on to other levels of training. Also, with young horses, you need to get them accustomed to having their hooves handled. Horses will naturally shy away from having someone lift their feet, which will make any farrier chores impossible if you have not done this training.

If you are selling colts or yearlings, this may be the extent of the training your customers will expect. After that, the next level of training would be breaking the horse to a saddle and bridle. This can be a time-consuming task but a necessary one in almost all cases unless you are only selling horses intended for heavy draft work rather than any kind of riding.

Beyond these basics, the required training will depend on how your horses are going to be used. As mentioned below, horses intended for sport or racing will require different training and upbringing than a horse intended for other purposes.

The specific techniques you can use to train your horses for all of these tasks are beyond the scope of this book. Professional horse trainers are available in many rural areas that can help you manage this part of horse raising until you have more experience to handle the training yourself. It does take some skill, along with a great deal of time and patience to properly train a horse.

Horses for Profit

The world of horse sales is far different than that of any other animal and should be researched fully before investing in any horse breeding operation. This is doubly true if you are raising horses to be sold as racehorses. Working horses or companion/ pet horses are a little more straightforward.

In either case, you are selling an animal whose value is a little harder to determine compared to an animal sold for meat. The value is based on *potential* performance, which is what you have to sell to your customers.

Another thing to consider is your customer base. Selling meat is usually a repeat business, whereas people are not going to buy horses from you at the same pace. This will mean you will have to look farther away for potential buyers, which makes transportation part of the transaction.

Regardless of the specific nature of the horse you are selling, your animals should have some basic training and be comfortable being handled by humans. Younger horses should be trained to walk with a halter and lead at the least, and older horses will need to have training that reflects the purpose for which they are being sold.

Selling sporting horses

Selling younger horses will mean you have to impress your buyers with their racing potential before the horse has ever entered a race. This is usually done by their pedigree. A horse born to past race winners will be of much greater interest than a horse with no such lineage. You can either buy into a bloodline by purchasing mares with good history or use stallions that also have a history of producing winning offspring.

Even with an excellent background, your horses will have to be healthy and in good physical shape in order to attract a buyer.

Keep a record of all the horses you have sold so you can further advertise that you have raised and trained several winning horses. Building a reputation is vitally important, so keep track of all your horses after they have sold.

Another option is to create your own line of winning horses by having them compete while under your ownership, as opposed to buying them. This can complicate your operation somewhat, but if you enjoy working with horses, it might not be that difficult to race or otherwise compete with your horses to develop a history of accomplishment.

Once you have a few ribbons to your horses' names, they will have more value, and you can sell these older horses rather than dealing only in untested yearlings.

Selling working horses

Working horses are less problematic to sell, as people are able to see their abilities demonstrated before they make the purchase. If you are selling yearlings, however, having a respectable bloodline will still be important to establish the potential of your young horse.

A horse sold as a companion will need to conform to your buyer's riding abilities and expectations. Some people will want a quiet and docile horse, but more experienced riders might prefer a horse with a little more attitude. You must be honest in your assessment of the horse so the buyer knows what he or she is committing to. Potential customers will want to take your horse for a "test drive," so make sure there is an open paddock or field available where they can ride undisturbed.

You will not need any special licensing to sell horses. Most transactions will include a clean bill of health from a veterinarian to ensure there are no underlying health issues with the horse.

Although horses are different from other livestock animals, they are still viable choices for running a farm. You might need more time and a larger initial investment than you would with other animals, but the profit potential is significant.

Chapter 8

Cattle

> ## Learn the Lingo
>
> - **Bull:** A male cow of any age
>
> - **Steer:** A male cow that has been castrated; typical in beef production
>
> - **Heifer:** A young female cow under 2 years old
>
> - **Cow:** A female that is older than 2 years or that has had at least one calf
>
> - **Calf:** A general term for any young cow
>
> - **Freshening:** The act of impregnating a cow to replenish her supply of milk
>
> - **Polled:** A cow that is naturally born without any horns

Cows are probably the most familiar of all livestock animals and a flexible one at that. You can raise them either to produce milk or meat. There is probably more information available for small-scale cattle farming than for any other type of animal, so you should not have any difficulties learning everything you need to know.

Pros and Cons of Raising Cattle

The main advantage to raising cows is their familiarity with customers. Cow's milk and beef products are popular, and you will not have difficulty finding consumers. Of course, with that familiarity also comes more competition, as there are so many beef products already on the market.

Keeping cows for milk will require a much heftier time commitment on your part because they will need to be milked at least once every day, if not twice. This must be done not only for efficiency or your profit margin but also for the health of your animals as well.

Cows are considerably larger than pigs, sheep, or goats, which will means you will need much more land and enclosure space. Handling them can be more of a challenge if you are on your own. However, they are generally docile creatures and not prone to bad behavior or difficulties with handling.

Housing and Pasture

Your housing needs will depend on whether you are raising cows for milk or meat. One meat cow will produce an average of 500 to 750 pounds of beef at slaughtering time, so even a small number of animals will give you a sizable amount of meat to sell. For a dairy operation, the average cow will produce about 10 gallons per day, and an exceptionally good milker can give you nearly 30 gallons per day. Knowing your animals' production potential can help you determine the size of your herd, and from there, you can design your barn.

Meat cattle will only require accommodations, but a dairy facility will also need space and equipment for milking, even if you plan to milk the animals by hand. *More information on this can be found in Chapter 12.*

A full-grown cow will require about 50 square feet of indoor space with access to a large pen or pasture space. If you are keeping bulls, you will need separate housing and pasture space for them in order to prevent any unwanted breeding. For female cows, you can either house them together in a large barn or separately in stalls as you would with horses. Individual stalls work better in a milking environment because you will need to isolate each cow while she is being milked each day, and it is much easier to do so when they are not all milling about in a huge barn. A cow needs a little peace and quiet in order to milk well.

Cows will do well on pasture, and you can greatly reduce your feeding needs if you have at least two acres per cow. You can use either conventional wire fencing or electric fencing; both work fine. With woven wire fencing, the posts will need to be sturdy, as cows can put a lot of weight into them if they choose to lean on a fence. Extra bracing around corners and gates is also a good idea.

On the other hand, they are not as likely to climb over or dig under, as some animals will. Your fence will not need to be higher than 5 feet, and most standard rolls of fencing are 54 inches high.

If you use electric fencing, you will only need two hot wires to enclose your cows. One should be at chest height, and the other should be halfway between that wire and the ground.

Breeds

The distinction between milk and meat breeds is clear. Cows that are bred for milk do not have the body mass suitable for meat, and cows bred for meat will not produce enough milk for commercial purposes.

Dairy cattle

With dairy cattle, be sure to research each breed and the type of milk they produce. Not all breeds produce exactly the same type of milk. Some will be heavier in fat content than others, which can be helpful if you intend to sell other milk products or have customers who want to make butter or cheese. You will also want to consider the volume of milk each breed will produce.

Unlike chickens that will produce eggs naturally, dairy cows do not naturally produce a large supply of milk all of the time. They only do so after they have just given birth to a calf, which means you will have to maintain a breeding program in order to maintain a relatively constant supply of milk. Once she has given birth and begun milking, she will continue to produce for you for up to a year until she eventually goes dry.

Growth or lactating hormones cannot be used to increase milk production on an organic farm, which means you will likely get a lower volume than conventional farms might with the same breeds. Keep this in mind when reading about the milk volume estimates of each breed, and use them for comparison only rather than as a strict expectation for production.

Holstein

Holsteins are extremely common in the commercial dairy business because they produce large volumes of milk — about 3,200 gallons per cycle. Their milk is lower in butterfat (3  percent) than most other cow's milk. These cows are a familiar sight, with their white and black splotched coat.

Guernsey

Originally a British breed, the Guernsey is a mid-sized to small cow that produces milk with high butterfat content (4.5 percent). They will produce roughly 1,700 gallons of milk for every milking cycle. They are a little bit less productive than other breeds, but they might be more suitable for a smaller dairy operation.

Jersey

The Jersey cow is the smallest of the milking breeds listed in this section, but it actually produces the most milk per pound of body weight. This makes them an efficient cow for a small-

scale farm. You will get 1,800 gallons of high butterfat milk (5 percent) for each milking cycle.

Ayrshire

Ayrshires are another breed well suited for more rugged surroundings; they came from Scotland originally. They will give you 2,000 gallons of milk with 4 percent butterfat.

Brown Swiss

This is another mid-sized breed similar to the Guernsey, though they produce more milk. Each cycle will give you 2,500 gallons of milk with about 4 percent butterfat. These cows originally came from Switzerland and are known to handle rugged terrain and harsher climates better than other breeds.

Beef cattle

Beef cows have the right muscular structure to provide the most meat per animal. With milk cows, you will be dealing with females, but male or female cows can be used for meat. Because females do have more value for future breeding, castrated males, or steers, make up the bulk of the meat cow population.

Although not usually raised for milk, you can get milk from a female beef cow once she has given birth to supplement your income streams. However, you will not get nearly as much milk as you would from a true milking breed.

Angus

Angus cows are known for their well-marbled meat, which is always in high demand. The cows are either sold red or black and are born with no horns, known as polled. They breed easily and calve well because they are small at birth. Because they are naturally hornless, cattle breeders often use Angus cows to introduce that trait into their bloodlines to produce hornless hybrids.

Hereford

These cows are larger than average, with broad forequarters and heavy muscling. Fortunately, they are docile animals that handle easily. They are red with a white face and come in polled and non-polled varieties.

Limousin

Like the Hereford, Limousin cat-tle are heavy and large-bodied. Originally a European breed from France, they are now pop-ular in North America for their lean meat. Limousins will grow fast, and they calve easily if you are breeding your own. They nor-mally have horns, but there are some polled varieties produced from crossbreeding.

Dexter

The black Dexter cow stands out as being one of the smallest breeds of cattle. They are typically half the size of a Hereford; they weigh in at fewer than 750 pounds. Although they are a less common breed in the meat industry, their size does give them an advantage for the small-scale farmer. They are often classed as a "dual purpose" breed because they also produce a good supply of rich milk, though at a lower rate than a milking breed.

Charolais

Charolais are white cows that grow quite large with heavily muscled frame. They will easily top 2,000 pounds in weight. These cows produce a lot of lean meat with a lower ratio of fat than many other breeds. They are not as popular as the Hereford but are gaining popularity.

Scotch Highland

Sometimes known as just Highland cattle, these cows are ideal for northern farms where cold weather is a concern. They have a long, shaggy coat and can adjust to extreme weather. The breed originally comes from Scotland. Their meat is lean, and they can thrive on poor pasture that other cows would not do well on.

Equipment

There will be considerable equipment required for a milking operation. *The specifics will be discussed in Chapter 12.*

Other than items used for milking, the main equipment will be feeders and waterers, which do not need to be particularly complex. A large hayrack for feeding and an open trough or stock tank for water work well for cows. Leads and harnesses will work when handling calves, and a simple harness kept on an older cow will make it easier to lead as well.

Feeding

The majority of a cow's diet will be either dried hay or grasses grown in the pasture. They are able to do well on poorer forage than horses because they have four stomachs. Their food is eaten, regurgitated, and chewed again to get all the nutrients. This is what happens when a cow is "chewing its cud."

A grass-based diet is considered the healthiest for cattle, though the practice of corn feeding is still common in large agricultural operations. A high-starch diet of corn will add fat to the cows faster and can increase the production of milk. This is not something an organic farmer would want to pursue, even though it technically does not violate organic standards in itself. A corn diet will seriously disrupt the digestive system and lead to many potential illnesses. Commercial feedlots use antibiotics to keep much of this in check, which is obviously not an option on an organic farm.

That said, corn is still a suitable addition to a grass-based diet and should not necessarily be abandoned altogether. It works as a treat and a motivator when moving cows into their milking stands.

If you plan to use your own pasture as the main source of food, you will need 2 to 3 acres of space per cow, or you can supplement with additional dried hay. Hay also makes a good food for over-wintering your cows, though you might want to add additional grain during the colder months. Timothy hay is a good quality choice. Corn, oats, barley, or wheat will make healthy grain options for a cow.

A pasture that is growing a mixture of native grasses is usually sufficient for cows without specifically seeding for any one type. Allowing one section of pasture to regrow while the other is being used is a practical way to maintain a steady supply of grass forage for your cows. Adding alfalfa or clover to the pasture growth is a good option to add a little extra protein.

Fresh water should always be available to your cows. A full-grown animal will drink more than 30 gallons a day, so be prepared to keep a sizable water supply available. A mineral salt lick is also a good idea to ensure a proper intake of trace minerals. These are large solid blocks made up of salt and other natural minerals cows can lick on their own to add these extra elements to their diets.

Breeding

Your specific breeding needs will be different depending on whether you are running a meat or milking operation. In a milking operation, you have to keep your milk cows lactating, so you will need to focus quite a bit more on breeding.

Breeding basics

The basic biology of cattle is the same for both meat and milking breeds. Their gestation period is just over nine months, and a heifer can first be bred around 14 to 16 months old.

Your cows will go into heat about every 21 days, and this is when they can be impregnated. Once the animal is in heat, she will only be receptive for eight to ten hours, so you must be diligent when watching your animals. Although this does create a rather narrow window for insemination, a bull will always know the right moment. As long as you have a bull present with her for the day of her heat, she likely will be impregnated.

As with horses, you will only want to keep your own bull if you have a large farm and intend to breed frequently. Renting an animal or taking part in artificial insemination are common options for small farmers. If using artificial insemination, you will need to be precise with timing, but professionals who offer this service will have the right expertise.

When a cow is in heat, other cows — even females — will try to mount her. Watch for mud on her sides and mussed up hair at the base of her tail as a sign that other cows have tried to mount her. There are small capsules on the market that attach to a cow's lower back that register when she has been mounted. These little plastic devices will break when pressure is applied from above.

In a milking environment, you will need to consider your supply of milk when it comes to breeding. Depending on how you time things, you might end up with a dry cow for several months. You might want to take this approach to give your cows a break between pregnancies, but your overall space constraints and budget might not allow realistically for too many nonproducing cows in your herd.

It is common practice to get your cows pregnant again about three months after they have had their most recent calf. This pattern roughly has your cow drying up around the time her next calf is born. If you wait until your cow is giving no more milk, you will have a nine-month wait for more milk.

You must also consider how to work the offspring into your breeding program. With meat cows, you usually are raising calves to replace slaughtered animals in your herd. However, in a milking operation, you will produce a lot more young animals than you really need, especially when the offspring are male. You can either sell the animals as meat or to other farmers. Either way, you should work this aspect into your milk operation so the animals are treated properly when you must dispose of them.

Calving

After about nine months of pregnancy, watch your cow for signs of the upcoming birth.

Your pregnant cow will lie down and stand up repeatedly and usually begin to pace around. This indicates the early stages of labor, which can last between four and six hours. During that time, you can move her to a clean stall or separate area of the barn so she will be not be disturbed when giving birth. Once she actively begins pushing, she will lie down and stay down. Once she begins to deliver the calf, the rest of the birthing should take between 30 minutes and two hours at the most. Calves are born with their front hooves out first so do not be surprised at that.

After the calf is born, the afterbirth will follow and should be shed within a few hours. It is not something you should try to remove yourself; let it naturally come out on its own. Do watch for it though because your cow can develop a serious infection if it is not delivered.

Until you have experience assisting with a birth, you should have a veterinarian present or nearby on call in case there are any complications. Cows generally will calve easily without any intervention from you, so there is no reason to expect problems. However, it is always a good idea to be cautious. This is especially true if it is your heifer's first pregnancy. *The next section describes how to handle the care of a newborn calf.*

Health

Cows might be large, but they are susceptible to many health issues just as smaller animals are. Watch over your herd carefully, and try to catch any possibly illnesses immediately. Most diseases are treatable, even without antibiotics.

Newborn health

When a calf first emerges from your cow, the sac of fluid surrounding it should burst so it can take its first breath. If that does not happen, you will have to break the sac yourself.

Once that is done, your cow will lick and clean the calf. You can help with a towel, but try to let the mother take care of the job. If the weather is cold, a towel is a better idea. The umbilical cord should fall off by itself. Disinfect the stump with a little iodine.

A calf will be up on its feet in no time and looking to nurse within the hour. First-time mothers might not be receptive to the idea. If she seems to be rejecting the calf, tie her with a short lead to keep her in place. After a short feeding, even a novice mother will accept her young.

If she truly refuses to feed the calf, you will have to bottle-feed. A colostrum replacement should be used for the first 24 hours before using standard cow's milk replacer. Colostrum is the first milk produced by a heifer, and it has more antibodies and protein than the milk produced after the first day.

Although the milk humans usually drink is cow's milk, it is not considered a good replacement when bottle-feeding a calf because it has been pasteurized. True milk replacer intended for bottle-feeding calves is a better choice. Cows will not require feedings as often as other animals, so you will only have to take care of this chore twice a day. Each feeding should be about 2 quarts of replacer.

Calves usually are weaned at 6 months old. Animals on pasture will begin to eat solids on their own, and a daily feed of sweet grain will help them wean. Eventually, the young should be separated from their mothers, which can cause some temporary distress on both parts. You still are able to milk your cow while she is feeding. She will produce milk in response to what is taken from her, so you do not have to worry about your calf going hungry because you are milking.

Male calves will need to be castrated unless you intend to keep them for further breeding. Males that are kept for beef production should be castrated, as it reduces aggression and has no effect on growth or the final meat product. You can hire a veterinarian to do this for you or tackle it yourself. Surgical procedures should not be attempted by a novice, but a technique using small elastic bands to cut off circulation to the testicles is something most small farmers can do themselves after observing a professional a few times. Castrate your male calves by 4 weeks old. The elastic band method usually is performed only on cattle because their anatomy allows for it. You cannot use this method with other large animals such as horses or pigs. Speak with your veterinarian before performing this procedure for the first time.

If your cows are a breed that develops horns, they should be dehorned for safety while they are young as well. Even well-mannered cows can cause injury or damage with their horns when they are older. This is a procedure usually done when cows are young and are just starting to develop their horns. Depending on the breed, this can be as early as 3 to 4 weeks old. A hot metal iron is held to the small horn buds to burn them off. Another option is a caustic paste, also used to burn off the area where the horns will grow. Both options can be painful to the calf, so it is best to do it as soon after the calf is born as possible. A trained professional or veterinarian should perform the procedure for you, though it can be done yourself if you have some experience.

Cattle diseases

Not all cattle diseases are found in all geographic regions, so ask your veterinarian which ones are common in your area. Vaccinations are a good idea with cows, as you can prevent many fatal diseases and protect your herd.

Scours

Scours is just the agricultural name for diarrhea in a young calf, and it can be fatal if severe or untreated. Calves that do not get a decent feeding of fresh colostrum at birth are more likely to develop scours, as they lack the antibodies to fight against the bacteria that can cause it. It can be caused by many different things, so focus more on the treatment than trying to pinpoint the specific source. You need to make sure the animal does not get dehydrated, which is the main cause of death from scours, not from the bacterial infection itself.

You can allow your calf to continue nursing or drinking milk, but also add regular doses of an electrolyte formula to help replace lost fluids. Electrolyte liquids are not medications but a liquid mixture that includes the proper blend of sodium and potassium to keep an animal hydrated. It is a better choice in these instances than plain water. Using this is fine for organic animals as long as the formula does not include additional medications. Antibiotics are a common treatment as well, which might be inevitable in a bad case. Rest and rehydration are usually sufficient.

Mastitis

This can affect dairy and meat cows, but it is a more significant problem for the dairy farmer. It is more common in dairy cows and will cause more economic losses. Mastitis is an infection within the tissue of the udder, which not only makes it painful for the cow to be milked, but it also spreads pus into the milk so it cannot be used. Symptoms include stringy or clotted milk and hard red teats. Antibiotics are used almost exclusively for treatment, which will eliminate that cow from producing organic milk, though any subsequent milk production after the mastitis has passed is fine for nonorganic sale.

Mastitis can be prevented with good milking hygiene and clean living conditions. Some cows might be more prone to it than others, and they should not be bred because the trait can be passed to its offspring.

Brucellosis

This illness, which is also called Bang's disease, is not a common disease in the United States and has been completely eradicated in Canada. Calves are usually vaccinated against it. An infected cow will have difficulty carrying calves to term, which leads to an increased number of spontaneous abortions. Any animals that are found to carry the disease are usually culled to prevent any further spreading. The only way for a cow to contract the illness is to be exposed to another infected cow.

Wooden tongue

As the name probably suggests, this disease affects the tongue and causes swelling and hardening. Cows with wooden tongue will drool excessively and work their jaws constantly due to discomfort. Small cuts inside the mouth from overly rough food can let the bacteria take hold. Eventually, the cow will be unable to eat and will die without treatment. There are no vaccinations for wooden tongue, but it can be treated organically with topical applications of iodine.

Black leg

This is another disease that attacks young calves and can be fatal. Although the symptoms include lameness and swelling in the legs, the infection starts when a calf ingests the bacteria in its food. This disease can kill a cow quickly, and there are no treatment options. Vaccinations are standard and should not be neglected.

Hardware disease

This is not a true disease but is well known among cow owners nonetheless. The name is just the common term for stomach problems caused by eating metal. Cows are nondiscriminating with what they eat and will eat nails, screws, small springs, or anything else that might be hidden in their pasture's grass. Sometimes, these things will cause no difficulties, but sharp objects can puncture the stomach walls. Short of surgery, there is little you can do, and it can be hard to diagnose. If you have a cow that is not eating, lethargic, and reluctant to walk around, suspect hardware disease.

The standard procedure is to actually feed your cow a smooth magnet designed specifically for this purpose. It will hold the object in place and help keep it from puncturing the stomach wall. If it is a cow destined for slaughter, you can retrieve the magnet then. A valuable cow might warrant surgery to remove the metal, especially if you intend to have it for several years.

Butchering

Even if you are keeping cows for dairy purposes, you might find yourself with older animals that are no longer useful for milking. At that point, slaughtering them for meat is usually the most efficient and cost-effective approach.

If you intend to do your own slaughtering, you will have the added responsibility of keeping a USDA-approved facility. Between that and the size of a cow, it is more practical to have the work done at a slaughterhouse. Only basic information is being presented in this section for proper killing techniques and the cuts of meat you can expect to find on a butchered cow.

Cows in a slaughterhouse are usually stunned unconscious with a blow to the temple, and then hung upside-down while their throat is cleanly cut.

Cuts of meat

Beef cows usually are not slaughtered until about 3 years old. Their weight at that point will vary by breed but will be about 1,000 pounds. As mentioned previously, you can expect between 500 and 750 pounds of usable meat from an animal that size. To put that in perspective, a whole cow can provide daily meat for a family of four for a year.

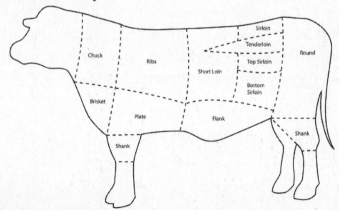

Depending on the local culture, there are different ways to butcher a cow. North American conventions are different from the way meat is cut in the United Kingdom, for example. When talking with your slaughterhouse, you should know all the proper terminology so you can ensure you get the cuts you are after. The following is a basic breakdown of the nine cuts you will get from a cow in North America:

- **Chuck:** This is the area around the shoulders that is used for some steaks and roasts. Some of this meat is sold as ground beef.

- **Brisket:** This meat is below the chuck and just above the base of the forelegs. It is a tougher cut of meat, which makes it ideal for barbecue cooking.

- **Shank:** This is the next cut below the brisket and is really the top of the foreleg. This is a cut for stewing meat.

- **Rib:** This meat is behind the chuck and makes up a large section of the back. Here you will find the short ribs, prime rib, and rib eye steaks.

- **Plate:** This is the section below the rib and around the belly area. There are more short ribs here, along with skirt steak.

- **Short loin:** This meat is farther along the back, behind the rib section. High-quality cuts such as T-bone and porterhouse steaks come from this section.

- **Sirloin:** This is the next section of the back, behind the short loin, known for more steaks.

- **Flank:** This meat is below both the short loin and the sirloin. You can get flank steak from this cut, but the meat is quite tough.

- **Round:** This is the hindquarters of the cow where the meat is actually lean. Eye of the round steaks and various roasts can be cut here.

You can get ground beef from any area, though the finer cuts of meat, such as the sirloin and short loin, are almost never used that way.

The exact amount of meat you will get depends on how the individual cuts are made. If you have your steaks cut thick, you will get fewer of them, for example. Overall, this is a general estimate of how many cuts of meat you will get per cow:

- 28 T-bone steaks
- 16 sirloin steaks
- 28 rib steaks
- 16 round steaks
- 28 roasts of various cuts
- Several packages of ribs
- Several packages of stewing beef
- 175 pounds of ground beef

One aspect of slaughtering cows is aging the meat, which involves leaving a beef carcass to hang in a refrigerated area to allow the connective tissues in the muscles to break down. This leads to tenderer meat, no matter what the cut. It is not necessary, but if you can work with a slaughterhouse that has the facilities to let your meat hang before butchering, you will have better quality meat that your customers will appreciate. Dry aged meat will lose a significant amount of overall weight due to moisture loss and is reserved for meat used in high-end restaurants and gourmet products. Wet aged meat is more typical. The meat is vacuum-sealed first so no moisture is lost. Dry aging can take a few weeks, but wet aging can be done in about four days to get the same level of tenderness.

Veal should be mentioned here. Although the term has some strong negative connotations, veal does not necessarily have to be connected to the poor treatment of animals. Strictly speaking, veal is meat from a young calf and does not *have* to include the tight restrictions on their movement. Veal is much tenderer than meat from an older cow, even if the calf has been allowed to run around the pastures during its short life.

Offering veal for sale is one way of handling male calves born in a dairy operation without letting animals go to waste. Selling veal can be a touchy subject, so be careful about labeling and letting your customers know the animals were not mistreated or abused.

Cattle for Profit

Large animals such as cows will provide you with a great deal of meat or milk, so their overall profit per animal potential is significant.

Selling meat

You can keep your operation more streamlined if you are able to sell your animals "in bulk." In other words, it is beneficial for customers to buy a whole, half, or quarter of a cow in each purchase. This approach not only means you have to manage fewer sales but also that you can avoid the legalities of selling meat. If your customers pick up their meat purchases directly at the slaughterhouse, you are not responsible for the cleanliness or certification of those killing facilities. You are selling animals, not meat.

Unlike a pig or a sheep, a cow represents far more meat than a customer will likely want in a single purchase. You might have difficulty finding one person who will want an entire cow's worth of beef. Halves and quarters are a more successful option, as long as you have enough buyers each time. You will need two customers at the same time to buy both halves or four customers to buy all four quarters. Otherwise, you will have to take delivery of the unpurchased sections, which will leave you with a need for a proper license to sell meat after all.

For a more flexible alternative, consider getting the license to sell individually wrapped cuts of meat. Keep in mind that a license to sell meat is not the same as a license to run a slaughterhouse. You will still need to outsource that aspect of the operation. Each state manages its own requirements for licenses, so check with your state's Department of Agriculture for the details. Slaughterhouses will have all the meat properly wrapped for storage, and you will only need the freezer space to keep it until it is sold.

Selling milk

Milk can be a harder product to manage, given its highly perishable and fluid nature. You will need the right equipment to hold and store the milk and keep it properly chilled until sale. You will also have to decide whether you have the time to hand milk or prefer to invest in milking machines for this job. *You can find additional details on this in Chapter 12.*

As with meat, you will need a license to sell milk, which states regulate as well. Some states allow you to sell unpasteurized milk, also called raw milk, while others do not. Avoiding the pasteurization process can save you time and the investment necessary to purchase the equipment, though not all customers will be interested in this product. Check on your local regulations before deciding which way to go.

With meat, you can sell in large quantities, such as a half cow, but milk is rarely sold in such a way. You will be dealing with much smaller orders and quantities of pints, quarts, or gallons.

Cows might be a little large for a first-time farmer, and they can provide you with more meat than you can sell until you have a customer base established. However, they will give you a reliable product that sells easily once you have the necessary experience. Remember, you will also need more space than when working with animals such as pigs or sheep.

CASE STUDY:
BEEF AND VEGETABLES
WORK WELL TOGETHER

Eric and Hannah Noel
Maplewood Organics
3550 Gore Rd.
Highgate, VT 05459
www.maplewoodorganics.biz

Maplewood Organics is mainly a beef operation, though they do also run an organic vegetable CSA program as well. Their family-owned farm spans more than 300 acres in Vermont, and they even run farm camps in the summer to help local school children (as well as adults) learn more about farming and where their food comes from. For folks who prefer to work, they have regular internships, too.

They are currently raising a 55-head herd of Galloway cattle for their beef production, as well as a sizable flock of laying hens for eggs. Three acres of their land has been dedicated to the vegetable garden. Maplewood's beef is sold mainly on a wholesale basis to stores in Burlington, though they have recently branched out to try Internet sales through their website. Beef is sold in individual cuts at the farmers market but also as large bulk packages or whole cows.

The Noels have a clear philosophy for their organic ways: "We chose organic because it flows with nature. To us, it is the path of least resistance. We choose not to fight nature, but instead learn from it and nurture it. It is more logical and life giving. We can produce more nutrient-dense food and increase the health of our bodies." They found the certification process to be a challenge but would not say it was difficult. Even just recently, they had to make changes to how they label their eggs because they had misunderstood the rules. As they say, you need to adjust your course to stay on target.

They have had little problem finding the right organic supplies for their farm. They grow all of their own feed and even use their surplus for neighboring organic farmers. Their veterinarian is conventionally trained but is agreeable to the organic way. Not that they need the vet that often — once or twice a year is all the medical treatment their animals require.

If you are thinking about starting a livestock farm, the Noels suggest that you take some time as an intern so you can get a proper feel for the work involved.

Chapter 9

Goats

> ## Learn the Lingo
>
> - **Buck:** A male goat; also called a billy
>
> - **Doe:** A female goat; also called a nanny
>
> - **Kid:** A young goat
>
> - **Wether:** A castrated male goat
>
> - **Chevon:** The proper term for goat meat
>
> - **Mutton:** A term also used for goat meat, though some regions use this term for meat from sheep
>
> - **Kidding:** To give birth
>
> - **Bleat:** The call or sound that a goat makes

Goats are a little less common than some other livestock animals, but they are growing in popularity and demand. You can keep goats for either their meat or their milk, much like a cow. They are fun animals with more personality than anything else you are likely to keep on your farm.

Pros and Cons of Raising Goats

A significant advantage with raising goats is their size. They are so much smaller than a cow or pig that they can be more manageable for smaller farm operations. Whether you are limited

in space or workers, goats will fit just about anywhere. Goats are also notorious for eating just about anything, which makes them ideal for areas with poor pasture.

Alternatively, they are less common in either the milk or meat markets, so your customer base will be smaller. The meat can be an acquired taste for those not used to it, though some ethnic markets are always looking for goat meat. Keeping goats fenced in can be another difficulty compared to other animals. Nothing can get through a fence better than a curious goat.

Goats are versatile animals that can produce a mix of incomes for a small farm. Milk, meat, and even high-quality fiber are all viable options for your business.

Housing and Pasture

Goats are seldom kept in an individual stall environment but rather housed together in large pens or open barns. They are not that different from other livestock and will need a secure place free from excessive draftiness.

If you are raising goats for milk, your barn will need to have a designated milking area with a stand to keep your goat restrained during milking. For space, each goat should have about 20 square feet of room inside the enclosure. If they have plenty of outdoor space, they might do well in closer quarters. For outdoor pasture space, you can house about ten goats per acre.

Goat house in a pasture

Goats are not grazing animals like cows but can be considered browsers. The difference is that goats will eat all kinds of brush as well as grass and weeds. This means they will make better use of your pasture, and you will seldom have to seed it to grow the right food. As long as nothing poisonous is growing, they will eat pretty much anything.

Keeping your goats inside the pasture can be a challenge. Young goats are small enough to squeeze through small openings, and older goats are curious and intelligent. They are known to open gates and find ways through most fencing. Traditional wire fencing can work, but it will need to be at least 5 feet high and have the wire openings much smaller so nosy goats do not get their heads stuck. There is fencing specifically designed for goats.

Alternatively, you can work with an electric fence system. It can work well for goats, but you will have to run more wires than you would with other animals. They can sneak between wires or go under them. Five or six wires are usually enough. The bottom wire should be about 6 inches off the ground, and the remaining wires should be spaced 6 inches apart, though the top two lines can be closer to 8 inches or a foot apart.

As with pigs, you can slaughter young goats in less than a year, which means you might want to organize your operation to avoid overwintering animals. In such an instance, you can accommodate your goats in a simple shed in the pasture. However, unlike pigs, a goat under a year old will represent a smaller amount of meat per animal. For customers looking specifically for kid meat, this can be a good approach.

Breeds

You can choose to raise breeds of goats that have been bred for higher milk production or those that have a larger body mass for meat. Given the state of commercial goat raising, most of the attention has been paid to breeding better milkers rather than improving meat breeds. Many people who raise goats for meat settle for one of the milking breeds because they are more readily available.

Dairy breeds

Dairy breeds are the most common among goats, and they can vary in production levels and butterfat content. Most milking goats are roughly the same size and have similar temperaments.

Saanen

The white Swiss Saanen is the largest breed of dairy goat, and the one that will provide the most milk. You can expect to get a gallon per day with a Saanen, and they produce nearly 3 percent  butterfat in their milk. Although they will weigh more than 150 pounds, they are docile enough to be around small children.

Nubian

Sometimes referred to as "Anglo Nubian goats," this breed is recognized by their long floppy ears. They do not produce as much milk as some of the other breeds, but their milk does rank highest in butterfat content at 5 percent, which makes it rich and flavorful. It also means their milk is favored for cheese making.

Toggenburg

The Toggenburg is a mid-sized to large goat that has distinctive white hocks and white stripes along its muzzle. Besides the striping, they can be any color. Their milk is low in butterfat at 2 to 3 percent, and their production level is average.

LaMancha

Although Nubians have long ears, LaManchas are known for the almost nonexistent ears. They will give you about three-quarters of a gallon of milk every day, with a typical fat content

of 4 percent. The breed originates in Spain though it was more thoroughly developed in America.

Meat breeds

A common practice with goat farmers is to raise a herd of dairy goats for their milk and sell the meat from male kids that are born. Most customers who purchase goat meat tend to prefer younger animals anyway, so it makes a good compromise. If you do not want to deal with the milking side of the business, you can find a couple breeds specifically intended for meat production.

Boer

This is originally a South African breed and definitely has a more muscular body shape than dairy breeds. Boers are always white with a brown head and have long ears similar to the Nubian. Boer does grow to 200 pounds,

and the bucks will reach between 250 and 300 pounds. Half of that weight will be saleable meat after slaughter if you allow them to grow to full size. Obviously, kids will be smaller.

Kiko

There is not yet a huge population of New Zealand's Kiko goats in North America, but their hardiness to adverse climates is making farmers interested in the breed. During cold weather, they will grow long coats for warmth. They are generally white goats, and males will grow a sizeable pair of horns if they are not removed while young. Although they do grow a long coat, they are not suitable for fleece or fiber purposes.

Myotonic

This is an unusual type of goat but has actually come to be a valuable meat breed. They are more commonly known as fainting goats because they literally fall over and go stiff when they are startled. This genetic trait is generally undesirable but has the odd side effect of adding considerable muscle tone because of the way they stiffen up when they "faint." Due to their habits, they might not be an ideal choice for a busy farm where the goats might get anxious or startled frequently.

Fiber breeds

It is worth mentioning some other goat breeds that are raised for their fur, or fiber. With a growing interest in handcrafted and natural products, there is also a growing market for organic fiber. Goat's hair can be spun into yarn, just like sheep's fleece. Shearing must be done annually or biannually, which makes the following breeds less work in that regard when compared to milking breeds.

With either of these breeds, you will only want to raise them if you have a reasonably clear pasture area. Trying to keep a herd of fiber goats on the same brushy field as a meat goat will leave you with a matted coat of fur filled with seeds and debris.

Angora

Angora goats produce a long coat, and the fiber is called "mohair." There are Angora goats in several different colors, and their coats are harvested by shearing once per year. They are not a hardy breed and will need a higher level of feed to supply the growth of their coats.

Cashmere

This is another breed that produces a woolly coat suitable for shearing, though you can also brush out a Cashmere goat to recover some of the fiber. The coarser outer hairs will need to be separated from the fluffy inner coat, which is the real cashmere fiber. It is mainly an Asian breed, but breeding stocks are available in North America.

When raising fiber goats, you do not necessarily have to learn how to spin. Many craftspeople are happy to buy raw fiber for their own spinning needs.

Equipment

There is not much equipment necessary for raising goats, but it does differ depending on the purpose of your herd. Dairy goats will need a place to be milked, and you will need a place to keep

Goat milking machine

all the associated equipment that comes with collecting and storing milk. *Chapter 12 describes this in more detail.* Meat goats will really only need a place to live and troughs for food and water.

If you are going to manage a fiber herd, you will also need combs, shears, and other woolgathering equipment. You can also outsource shearing to a professional, which can eliminate your need for many of the necessary tools.

For taking care of goat hooves, you should have a proper hoof knife or hoof trimmers. If your pasture has rocks or rough areas, a goat's hooves will naturally wear down, and this will not be something you need to deal with. A professional farrier also might be a better option until you have developed some experience in handling your goats.

Feeding

Goats are tolerant of nearly all types of food and can thrive on pasture that no other animal can survive on. However, do not assume you can neglect to properly feed your goats for them to stay healthy.

If you are raising more than one type of animal on your farm, your goats can share pasture with larger animals as long as the field has more than just grasses growing. Goats do not just graze on grass like horses or cows. They will need a mixture of rougher forage in order to be happy and nutritionally healthy. That includes branches, brambles, and any tree bark they can get their teeth into. Other animals will graze around trees, but goats will strip them down as far up as they can reach, so do not expect any small trees in a goat pasture to survive very long. A ring of fencing around larger trees is a good idea to keep your goats from eating the bark.

When not relying on pasture alone, you can feed your goats dried hay and some grains. Good quality hay, such as Timothy or alfalfa, is great for goats and is usually given out in two or more feedings per day. Grains such as oats, wheat, corn, or barley are all healthy additions to a goat's diet but only should be considered a supplement to pasture or hay. Milking does can be lured into the milking stand with a little extra grain when they consider it a treat.

To provide enough protein for goats, their feed should include some legumes, such as clover, as well as soybean or cottonseed meal. Root vegetables are another choice to add variety and additional nutrients to a goat's diet. You can buy organic commercial goat feed, though it is a less commonly found product than feeds for some other animals. Using premixed commercial feed would save you the added trouble of trying to balance protein and minerals if you are feeding mostly dried hay.

Goats can drink up to 2 gallons of clean water per day, so you must have a large enough watering tank to give them a plentiful supply. Mineral blocks or solid salt licks are a good idea for goats, too. It is an easy way for you to supply necessary minerals and salt into your animals' diets. They will lick the blocks when they need to, as their instincts will prompt them to find salt when they need minerals.

Breeding

Meat goats are bred to produce more animals as you continually slaughter some of your herd. As with cows, a milking farm needs to have a continual breeding program so the milk supplies do not dry up.

Breeding basics

Does are able to be bred at about 8 months old, though Boer goats mature quickly and reach breeding age at just 5 months old. Their weight is more important than the exact age. A young doe should be about 85 pounds before she is bred for the first time.

Their reproductive cycles are about 21 days long, which means you will have a chance to get your doe pregnant every 21 days. During that period, a window of 24 to 48 hours is available when she is receptive and in heat. To complicate things, not all goats will have this cycle continually through the year. In climates with a cold winter season, a doe will usually have her heat cycles during the fall and winter.

Does that are in heat will show the typical signs, including restlessness, increased vocalizations (bleating), and mounting other does.

Keeping a buck on the premises will require a separate pasture and housing, but they are not as difficult to keep as a stallion would be. Even so, most people do avoid keeping bucks around unless they have a frequent breeding program in place. Male goats also have a tendency to be aromatic, which adds to their undesirability around the farm. You can either rent a buck when you need one or have your goats artificially inseminated. Artificial insemination is an acceptable practice for breeding animals within organic standards, as it is not a chemical procedure in any way.

Once pregnant, a doe will kid in about 150 days, or about five months. Having twins or triplets is quite common, so plan for the likelihood of more than one offspring with each breeding. You are more likely to get a single kid if this is your doe's first pregnancy. After that, you can almost certainly expect at least twins each time. If you do have a doe that consistently has single births, do not continue to breed her unless you specifically want that trait to be passed on.

After she has kidded, it is customary to let a doe rest while she cares for her young before rebreeding her. A period of two to three months is the usual time frame. This means she will stay in milk almost continually, and you will not have to house and feed a nonproducing dry goat. On the other hand, because pregnancy only lasts five months, you might choose to wait until she dries up naturally, which can be ten months or longer before having her impregnated again, as this is not an excessive amount of time to handle a dry goat.

Birthing

Because most goats get pregnant in the fall, you can expect most of your kidding to happen in the spring.

When she is ready to begin her labor, a doe will usually separate herself from the herd and try to be alone. She will pace around and bleat softly. Move her to a clean stall when you expect the birth to be imminent. Most does will lie down to kid, but some might choose to stand. Do not try to restrain her or force her to lie down for your convenience.

If you are completely unfamiliar with goat birthing, have an experienced assistant or veterinarian around for the event. Once you have been through this a few times, it should require little outside help. With their smaller size, you are going to be more able to assist with a goat birth compared to a cow's. A little help when kid is stuck or positioned poorly can make a big difference — as long as you know what you are doing. Goats will be born nose and front hooves first. A breech kid (coming out backwards) can be turned but that type of assistance should be left to a professional if possible.

Once she starts to push, the first kid should be out within an hour. If it is taking longer, contact your veterinarian. After the first kid is born, watch for the afterbirth to be expelled. If she still seems to be laboring with no afterbirth, expect another kid on the way. Once the afterbirth is out, your doe is finished kidding.

The doe will lick and clean her kids, but you can always gently assist with a dry towel. This is a good idea if there are two or more kids for her to care for. Within an hour, they should be on their feet and nursing. *The next section describes how to handle the care of a newborn goat.*

Health

Goats have a reputation for being sturdy animals, probably because they are known to eat just about anything and enjoy rougher terrain than other livestock. However, they are susceptible to many diseases, usually ones similar to those you would find with sheep or cows.

Newborn health

As with other livestock animals, a doe may refuse to feed her newborn kid. This can pose a problem because you are more likely to have two or more kids to care for if the mother rejects them. Newborn goats naturally need small amounts of milk at a time and will feed every couple of hours. So, if you are bottle-feeding with colostrum and then a goat's milk replacer (as mentioned in the section on newborn cows), you will have to do the same thing. For the first week, your kids will take around 2 ounces of milk every two to three hours — even during the night. By the end of the first week, they will start to drink larger amounts, and you can stretch the feeding times out to every four hours. As the kids need more milk each time, you can wait longer until the next feeding. Eventually, they will need 15 to 20 ounces of milk at three feedings per day.

Whether you are bottle-feeding or your kids are feeding naturally with your does, you can start weaning them onto hay and grain at about 2 to 3 months old.

Because the young goats tend to feed so often, plan on leaving your kids with the does rather than try to separate them between feedings. When they are ready for weaning, you can start keeping them apart to encourage your young kids to eat hay and grass.

While young goats are still on milk, you will need to watch them for diarrhea. Most farmers call it scours, and it can be a serious problem with young animals. It can be caused by a bacterial problem or even just indigestion from eating too much grain or hay. Call your vet right away. Dehydration can kill a kid within a day or two if left untended. With a little experience, you can treat your own goats with electrolyte formulas — even over-the-counter mixes you would use for children.

Goat diseases

Many of the diseases to which goats are susceptible are the same as those already listed for other animals, though a few of them are unique to goats.

Mastitis

As with cows, female goats can develop mastitis in their udders that will affect milk production. That makes it a serious disease for anyone managing a goat dairy. The symptoms include sore red teats (which may not spread to the entire udder) and clotted milk. The infection will create stringy pus that will get into the milk, which makes it unusable. You can usually prevent mastitis from developing in your goats with clean stalls and hygienic milking habits. It is not generally fatal if treated; however, the antibiotics necessary to clear it up will affect your goat's organic standing. Once treated, the milk is fine to drink for any nonorganic sales from then on.

Ketosis

This condition is also called acetonemia and is most commonly seen in pregnant does. The goat's body is not getting enough nutrients from her feed, and it will start to metabolize her fat reserves. All does should have good quality hay and extra grain when they are pregnant, especially if you know they are carrying multiple kids. The symptoms of ketosis include lethargy, stiff-legged gait, and a distinctive sweet smell to the animal's breath. If you suspect your doe has ketosis, contact a vet, though you may be able to treat it yourself once you recognize the signs. High-carb supplements such as molasses or Karo syrup can be administered if the animal has enough of an appetite to take it. Your vet can also provide glucose treatment. None of these treatments will have an effect on your animal's organic certification because the treatment is nutritional rather than chemical.

Blackleg

Blackleg is another disease you may encounter with cows or sheep as well as goats. It is fatal, and you should plan to get all of your goats vaccinated for it because there is no treatment if it is contracted. Bacteria and spores live naturally in the soil, so there is a potential for it even when no contact is made with other infected animals. Your young animals will develop lameness and swollen legs, and they will typically die soon after.

Caprine arthritis encephalitis

Animals affected by caprine arthritis encephalitis (CAE) will have arthritis-related symptoms such as stiff joints. Infected goats may have swollen knees and difficulty walking. Not all goats with CAE will show symptoms, which can make it hard to detect in a herd. There is no treatment for it, but it is not usually fatal — in fact, it is not that much of a problem with meat goats at all. But goats with CAE can have reduced milk output, which makes it a problem for dairy herds. When you have animals with CAE, slaughter them to prevent further spreading of the disease.

Foot and mouth disease

Foot and mouth disease (FMD) can be a major problem for live-stock farmers, though it can only infect your animals if they have contact with other ill animals. That includes having the bacteria come onto your farm through any soil on your shoes or wheels if you are visiting another farm with possibly ill animals. If you know of an outbreak in your area, you can get disinfectant to dip your shoes in as you enter your property to help keep FMD from spreading. Symptoms include sores on the nose and around the mouth and on your goats' hooves. It is fatal and cannot be treated. Have your animals vaccinated twice a year for protection.

Butchering

Goats possess an advantage over cows and pigs at slaughtering time because of their smaller size. Although slaughtering and butchering a cow is too big a job for one person, a goat can be done without too much difficulty. If you do your own slaughtering, you will need to maintain a USDA-approved facility. Which route you choose take will depend on how much red tape you want to deal with on your farm. The information in this section assumes that you will be using a slaughterhouse and will not describe specific butchering techniques.

Cuts of meat

The tenderest meat from a goat comes from a young kid, between 3 and 5 months old. Meat from kids at this age is called "cabrito" and can sell better than meat from older goats. The animals will weigh between 25 and 50 pounds, which will leave you with 10 to 20 pounds of meat after slaughter. If you prefer to get more meat per animal, let them grow for nearly a year before slaughtering. It will not be as tender as the cabrito though.

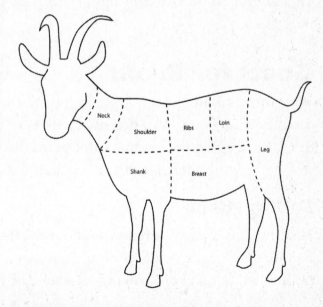

Goats are cut similarly to cows, though they are not divided into so many specific cuts or types of meat. Because goat meat is not as common as beef in North America, people are less familiar with the small divisions of the cuts. Goat meat tends to be sold as larger pieces with simpler terminology:

- **Forequarters:** This is the front portion of the animal, including the fore shoulders.

- **Ribs:** These are the sides of the goat, toward the front. Ribs and rib chops are found here.

- **Loins:** These are the sides, but they are located a little farther back toward the hindquarters. Loin chops are a popular cut.

- **Leg cuts:** These are the back legs of the goat. Roasts, steaks, and stew meat can come from the leg area.

You can have your meat aged to increase its tenderness — usually around two to three days is sufficient to improve the texture. This would be particularly important with older goats, whose meat is usually tough.

Goats for Profit

With three different ways to earn money from your goats, you can maximize your profit potential with even a small herd of animals. By keeping breeds that offer more than one way to make money, keeping a flexible farming plan is easy.

Selling meat

Goat meat is not as popular as other meats, so you will have a more challenging time finding customers, though the meat's popularity is quickly growing. Because of this limitation, the typical route of selling half or quarter animals may not be suitable. Of course, a smaller goat will not provide as much meat, so some customers may be fine with a larger portion, such as a half animal. You may be more successful selling smaller individual cuts instead, as most people do not eat goat meat as often as they would beef or pork. If you have access to certain ethnic communities where goat is more popular, that can work to your advantage.

Selling smaller cuts of meat will mean you have to have a meat-sellers license for your farm. Check with your state's Department of Agriculture for the specific details and how to apply, as the process and fees vary from state to state. Selling goat meat usually falls in the same category as beef or pork, rather than exotic game meat.

Selling milk

Goat's milk is another up-and-coming niche product that will not have the same competition as cow's milk. Some people with various levels of milk and lactose allergies will prefer goat's milk, so there is certainly a demand for it. It is also a popular product to sell to soapmakers who specialize in goat's milk soaps. In such cases, you can sell milk without a license as long as it is marked "not for human consumption."

Handling and processing milk is a large topic in itself. *This topic will be covered further in Chapter 12.* You will need to have the right equipment to chill and store the milk, and you may even need milking machines if you have a large enough herd to warrant such an investment.

State regulations will vary on the sale of milk, and you may be required to pasteurize your milk in order to sell it legally. Either way, you will need to apply for a proper license in order to sell your goat milk. Your own state will regulate the licensing, and your Department of Agriculture should have the information you need.

Selling fiber

As mentioned in the section on breeds, another product you can earn money from is your goat's hair or fur. The crafting community is a potential customer base for raw goat angora or cashmere. You can sell the raw material right off the goat to those who like to card and spin their own wool, or do some of the processing yourself and sell it. Carding refers to the process where the raw fiber is pulled between two combs or brushes to smooth and straighten the fibers. Once carded, the fiber can be spun into yarn. Rather than attempt to do all these steps yourself, you can offer your fleeces at whatever stage you prefer. Fiber that has been carded but not spun is called a roving, which is a good way to offer your fiber for sale.

Overall, goats are a fun animal to raise on a farm, and they do have some great profit potential. However, due to their less popular status for milk and meat, you may find it a little harder to develop a customer base. If you prefer to work in a smaller niche, goats may be ideal.

CASE STUDY: MAKING A BUSINESS WITH JUST ONE PRODUCT

Nancy Nathanya Coonridge
Coonridge Organic Goat Cheese
47 Coonridge Dairy
Pie Town, NM 87827
www.coonridgegoatcheese.com

You do not need lush pasture to make your farm business a success. Nancy Coonridge has a thriving goat cheese business in a very remote patch of dry New Mexico. Her herd of 60 to 70 does is doing just fine, and she is selling her gourmet goat cheese directly to her customers without any middleman. Coonridge has a mixed herd of Alpine, Nubian, LaMancha, and Oberhasli goats, and they run free-range without fences.

Her animals are kept only for milk production, and cheese is her sole product. Not only are her animals organic, but all the ingredients that go into her herbed cheese are as well. She originally began with 30 does and did all the milking by hand. Driving to the next large city for sales took three hours, so Coonridge worried about making enough sales to make the trip worthwhile.

She has been certified as organic since 1998. The only difficulty in staying organic she has experienced has been finding organic hay suppliers in her region.

Coonridge shares her cheese-making skills by hosting workshops at her farm. She offers suggestions for other potential farmers. "Sometimes new businesses are told to have an exit plan in case things do not work out," she says. "I say, find something you are passionate about and MAKE it work!"

If she could go back and do things differently, she would have planted fruit trees when she first began to raise her goats. She also thinks her herd would be easier to manage had she stuck with one breed of goat. Even so, all the choices she has made have been a valuable education. Coonridge is happy with the size of her operation these days and does not plan to expand.

Chapter 10

Sheep

Learn the Lingo

- **Ram:** A male sheep of breeding age

- **Ewe:** A female sheep of breeding age

- **Lambs:** Young sheep

- **Wethers:** Males that have been castrated

- **Mutton:** Meat from sheep over 1 year old

- **Lambing:** To give birth

- **Polled:** Naturally having no horns

Sheep are best known as livestock animals kept for their fleece, but they are just as versatile as goats. You can raise sheep for meat and even for milk if you wish, which gives you plenty of options for your animals.

Pros and Cons of Raising Sheep

Along with goats, sheep are versatile and can be raised for several different income-generating options. They are also a much more manageable size compared to larger animals such as pigs or cows. Like goats, sheep will commonly have twins when they give birth, which means you will naturally get a larger return on your investment if you are breeding your animals.

One potential negative to raising sheep is their woolly fleece. A sheep's coat can be a bit of a nuisance. The wool can get caught on some kinds of fencing, particularly barbed wire, and can shelter a number of worms or insect pests. Regular shearing will make all the difference in this, no matter the purpose for which you are raising your sheep.

Your potential customers for sheep's meat or milk will be fewer than with other animals because they are not as well-known food products. But just like with any other lesser-known items, this also means a lower level of competition from other farmers.

Housing and Pasture

The housing needs for sheep are similar to those for goats. Inside a barn, you will want to allot at least 20 square feet for each sheep. For pasture, you can expect about ten sheep to do just fine on each acre of land.

Many breeds of sheep will grow to slaughtering size within one year, so you can simplify your housing if you only raise your animals through the summer and fall. By not keeping the animals over the winter, you can use a simpler enclosure rather than a full barn.

Barbed wire fencing is not ideal for any kind of livestock but should be avoided particularly when raising sheep. The barbs will almost certainly snag their wool, which can lead to injuries if the animal gets tangled and cannot pull free.

A sheep with a full fleece will be well insulated against the shock of an electric fence, but if your sheep have been trained to respect the fence line, it is unlikely they will be testing the fence as they get older. And as long as their noses are still able to feel a shock, you should not have any difficulties using electric fencing with a herd of sheep. A setup similar to that constructed for goats, about five wires of fence, is also perfect for sheep.

Breeds

When choosing a breed of sheep for your farm, the lines between wool, milk, or meat breeds are blurry. Unlike cows, which are clearly broken into groups of dairy or meat cows, sheep raised for meat are usually good to raise for wool. Many sheep also are good milkers, so the need to get the perfect breed is not as much of an issue. Generally, larger animals will provide more meat and more wool. Sheep's milk is not a major commercial product, and there are not many breeds developed strictly for that use.

For a farm operation that is going to use the sheep mainly for wool production, a white-fleeced breed is the best choice, though some crafters are starting to develop an interest in naturally colored fleeces rather than white wool that has been dyed.

Dorset

Dorsets are excellent sheep for the new farmer as they grow quickly, give birth easily, and frequently produce twins. The rams will grow large curved horns, but you can find Dorsets that are polled as well. These sheep are white and can produce about a 15-pound fleece. At maturity, they are average-sized animals with rams weighing roughly 225 pounds and ewes being about 50 pounds lighter.

Cheviot

A Cheviot sheep is also white but has black legs and nose. They will mature to a smaller size than most other breeds and consequently produce less fleece — only about 7 pounds. A Cheviot ram  will grow to approximately 150 pounds. Cheviots grow slowly, so they will not suit any farming arrangement where you intend to slaughter the lambs during the first year.

Montadale

The Montadale was developed as a breed in the United States from the Cheviot and the Columbian breeds. These are large sheep that will give you a heavy fleece — more than 10 pounds — and the rams will reach 250 pounds when fully grown. Like Cheviots and Dorsets, Montadale sheep have white fleeces.

Oxford

If you are looking for a large animal, choose to breed Oxfords. These are some of the largest sheep, with rams weighing upward of 300 pounds and ewes close to 200 pounds. They are usually kept ex-  clusively as meat animals, though they will provide a decent fleece if you want the fiber as well. Oxfords have grey or brown fleece with darker brown to black legs.

Hampshire

This is another average-sized breed, originally from England, that will give you about 8 pounds of fleece at shearing time. They are white animals with black faces and legs. Unlike the Cheviot breed, the Hampshire will grow rapidly and makes a good candidate for a summer-only raising schedule.

Suffolk

These are large-bodied sheep like the Oxford but have white fleeces and black faces. The ewes are good mothers that often birth twins and will

provide a good supply of milk. An average fleece from a Suffolk will be about 8 pounds.

Merino

You want this breed if you have a wool farm. Their fleece is highly prized because it is softer and finer than most others. Merinos will be smaller than average, so they are not as ideal for meat purposes. Some lines of Merino are polled, and some have large horns that will curl close to the head. You can get between 15 and 20 pounds of fleece per shearing with a Merino.

Katahdin

This breed deserves inclusion because it has a unique shedding fleece that does not require any shearing. This will be of no use to a wool farmer, but if you want to raise sheep for meat or milk without the hassle of annual shearings, this may be the ideal breed for you. They are a heavy-bodied breed, most often kept for meat purposes.

East Friesian

Not many sheep breeds are bred specifically for dairy purposes, but the East Friesian sheep is one of them. You will get about 100 gallons of milk per breeding cycle (the time between giving birth and when  the ewe stops producing milk). East Friesians will also produce about 10 pounds of white fleece each year.

Equipment

There are more equipment requirements with sheep than most other livestock animals because of their heavy fleece. Most of your needs will depend on the purpose of your flock. Animals raised for meat may not need shearing if you plan on slaughtering them after just one year. You also can avoid the whole shearing issue if you hire someone to come to your farm and shear your sheep for you. It is a once or twice a year chore that many small farms can contract out to avoid the time, effort, and equipment costs.

Compared to slaughtering, shearing sheep is not as big or as disagreeable a chore. If you choose to do it yourself, you will need a good set of electric clippers. It is not a huge expense, but you will have to replace the cutting blades frequently to keep them sharp. *You can find more on sheep shearing in the Wool section of this chapter.*

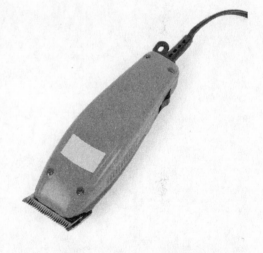

For herds that are kept for milking, you will need the most equipment, whether you just need tanks for refrigerating the milk or large milking machines. *Go to Chapter 12 for more information on managing a home dairy.*

Aside from these areas, there is not too much you will need when raising sheep. Troughs or barrels for food and fresh water are essential, though the specific designs are not that important.

Feed

Sheep are similar to goats and can thrive on pasture, though they are not quite as adept as goats at eating nearly anything that grows. An open pasture that grows wild will do well for sheep without the extra work of having to reseed it with certain plants. During the summer months, your fields should provide all the food your sheep require. A little added grain is a good addition, but only as a treat or as a lure to get your animals to follow you. Oats, corn, and barley are good for sheep.

During winter months or in early spring before the grass has begun to grow, you will need to feed your sheep dried hay as the

main component of their diet. Additional grain during the winters is helpful to build a little extra fat for protection against the cold. When there is no access to pasture, feed your sheep about 6 pounds of hay and 1 pound of grain each day.

Sheep do not need a lot of extra protein because their stomachs will produce most of what they need from forage. During the winter or when a ewe is pregnant, you may want to add some protein-rich foods to their diets. Alfalfa and clover hay or soybeans are two suitable additions.

On average, a sheep will need 1 ½ gallons of fresh water every day to drink, so plan your watering capabilities to accommodate. To make sure they have the right amounts of trace minerals and salt, provide them with a salt lick block specially formulated for sheep, which you can find online or at most livestock supply stores.

Breeding

The simplest breeding program is to raise your lambs until the fall and then slaughter the entire herd for meat. This requires no breeding, as you just buy new lambs each year. Obviously, this is not a suitable approach if you are running a dairy or wool operation or would rather raise your own lambs rather than buy them. So, in those cases, you will want to know more about handling the breeding.

Breeding basics

Deciding when a ewe can first be bred is a difficult question. They typically become sexually mature around 8 months old, but the time of year they were born will play a role. Sheep generally hit puberty in the fall of their first year, within reason. A lamb born earlier in the year will be a little older and larger at puberty than one born later. If your young ewe is less than 90 pounds in weight when she comes into her first heat, wait a few months until she reaches that size.

The estrus cycle for a ewe is between 13 and 19 days long and usually averages about 17 days. During that period, you will have roughly 24 hours in which to have her impregnated by a ram (or by artificial insemination). The timing can be difficult and made more of a challenge in that ewes rarely show any outward signs of being in heat. They will sometimes pursue a male if there is one nearby and even mount a ram if given the chance. You can always just leave a ram in with your females for two or three months during breeding season to ensure the most pregnancies, but you will lose a little more control over your breeding, and you will not have the precise dates for expected lambings.

Part of a sheep's estrus cycle is tied to the seasons, but that will depend on the particular breed of sheep you are raising. Some sheep can be bred year-round, such as the Dorset. Otherwise, a sheep only will go into heat when the shorter day length of fall triggers her cycle. You will usually have several months to work with, and the exact timing will vary greatly by your geographic location. You generally can breed sheep during the fall and early winter months, which means lambing season will be spring and early summer.

Once pregnant, it will be between 144 and 152 days until the ewe gives birth, or roughly five months. Twins are quite common with sheep, and you may be lucky enough to get triplets on occasion.

Because ewes are not likely to go into heat during the spring or summer, you have more flexibility on how to handle a ram in your flock. If you want to have a ram as part of your operations, you can let him run with your ewes during the spring and summer without having to worry about unplanned pregnancies. You do not often get this option with livestock animals.

Once your ewes have lambed, you will need to determine when to breed them again. If you are raising your sheep for milk, you must plan the pregnancies to keep your ewes providing milk without too much of a disruption once her lambs are weaned. Waiting about three months after lambing before impregnating the ewes again is routine if you are looking to maximize your lamb or milk supplies.

Lambing

As lambing time approaches, do a little extra shearing around your ewe's udder so your newborn lambs will have an easier time getting milk. If your sheep were recently sheared, this may not be necessary. Having clean and unobstructed udders is important as your lambs are born.

You will find that sheep birthing is much like goat birthing and can be handled without a vet if you have a little experience. When your ewe is ready to give birth, she will move away from the other animals in the pasture and start to scrape at the ground with her hooves. She may also lie down when ready to lamb as well. When you see her doing these things, bring her to a clean stall or area of the barn so you can keep her safe during the birth.

You should see a nose first, along with the two front hooves as each lamb is born. If you see hooves only, gently reach in and see where the nose is. Call a vet if you think the lamb is going to be breech. With more experience and some patience, even a breech birth eventually can be handled on your own. Sheep do not usually require much assistance at birthing time.

While the ewe is pregnant, you can have a vet use an ultrasound to see how many lambs she is carrying. If you have not done this, watch for the afterbirth. Once it is expelled, there will be no more lambs. If the ewe is still laboring after the first lamb is born, you can expect another lamb.

Dry and clean off the first lambs because the ewe may not do so right away. Keep your new mother and young separate from your herd for about two days before letting them back into the pasture with the rest of your animals. *More on how to care for a newborn lamb is in the next section.*

Health

Sheep are much like goats and have many of the same health concerns. Always be on the lookout for potential symptoms so you can treat or isolate your animals as soon as an illness crops up.

Newborn health

A young lamb should start to nurse after about an hour. First-time mothers may not take to their lambs right away, so be patient. They are also more likely than experienced mothers to ignore their young altogether, so do not leave your new lambs alone until you see that the ewe has taken to them.

When a ewe dies or refuses to care for its young, you will have to bottle-feed. These orphaned lambs are sometimes called "bum lambs" or even just "bummers." Use a milk replacer formulated for lambs to get the right amount of fat in the milk. Newborn lambs feed in small amounts frequently, so you will have to adapt to that instinct if you are going to bottle feed. For the first few days, plan to have a bottle ready every two hours. During the night, you can stretch that to around three hours between feedings. Fortunately, you will not have to keep this pace up for long. After about three days, you can give larger feedings every four hours. After a week, space out the bottle feedings to every six hours. At this point, the lambs should be taking about ½ cup of milk replacer at each feeding. If you give them a feeding late at night, they will usually be fine until morning. At a month old, you should be feeding them twice a day.

You can start to wean your lambs at about 6 weeks old or even earlier if they are showing more interest in eating grass and hay on their own.

Tail docking new lambs is one issue that is unique with sheep. Some breeds of sheep have naturally long tails that will get heavy with wool as their fleece grows. The tail fleece can get contaminated with feces and dirt, which leads to potential fly infestation. For sheep raised for their wool, it means you will get less usable (clean) fleece from each shearing, so it is common for commercial wool farms to dock the tails of their animals soon after they are born. This means removing most of the tail by various means. The most humane method is to attach a tight rubber band to the base of the tail when they are about a week old. Loss of blood flow causes the tail to die, and it will drop off on its own. Do not attempt this on your own — get a veterinarian's assistance.

You can decide if you want to dock tails, as it is not a necessity, particularly if you are raising your sheep for meat. You may want to try docking some of your lambs to see if it adds any ease to shearing or improves to fleece quality. With short-tailed European breeds, you will avoid the issue completely.

For young male lambs, castration is another potential chore unless you are planning on slaughtering them while still lambs (under 6 months old). Because lamb is more popular meat than mutton, this is a definite possibility. Intact males will grow more quickly, so it would be advantageous to avoid castration. Any males that are to be kept to adulthood but not for breeding will need to be castrated, or they will grow into more aggressive rams. Castrated males can be kept with ewes without fear of un-planned pregnancies.

Sheep diseases

A large-animal veterinarian is an asset for any livestock farm, but you can help protect the health and well-being of your animals if you are able to detect diseases or illnesses as quickly as possible.

Worms

Worms are technically not a disease, but they are worth mention-ing because sheep are more prone to getting them than any other kind of livestock. The reason is that sheep will graze much closer to the ground than other animals and will ingest more worm lar-vae. Various tapeworms or roundworms will infest the digestive system, which leads to poor health, slow growth, and possible death. The standard chemical treatments for worms cannot be used with organic animals, so you will need some alternatives at hand when raising sheep. You can use garlic or diatomaceous earth to help treat worms organically. Another technique is to ro-tate your pasture space. Worm eggs can survive in the soil for several months, which then end up being redigested as the ani-mals are grazing. If you move your animals to a different field, the eggs will eventually die. Three weeks is typically long enough to kill off any larvae in the grass.

Mastitis

Even if you are not keeping sheep for milking purposes, you need to be on the lookout for mastitis. It is an infection of the udder that will pass clotted pus and occasionally blood into the milk. Sheep that have developed mastitis will not be able to feed their young, and you certainly will not be able to milk them for saleable milk. The teats will be red, warm to the touch, and some-times even hard, but the full udder may not be affected. It may be a bit of a struggle to get the ewe to let you milk her, but you should do so to help relieve the pressure from accumulated milk.

The infection is caused by dirty conditions, especially with milking procedures. That can mean manure in the milking area, dirty hands when you milk, or soured milk left on your equipment. As long as you use clean tools and wash your hands as well as your animal's udders at milking time, it is unlikely that your ewes will have a problem with mastitis. The only treatment is antibiotics, which will render your sheep and her milk nonorganic. Once the infection is gone, the milk will be suitable for drinking again.

Ketosis

Similar to the same condition in goats, ketosis is a potential problem with pregnant ewes, especially those carrying twins or triplets. Ketosis occurs when the drain on the body becomes too great for the nutrients being taken in, and fat begins to be metabolized. This can drastically alter the sugar balance in the blood and cause lethargy and difficulty walking. When you have pregnant ewes, add extra grain to their diets and have top-quality hay available for feeding. Ewes tend to be pregnant during the winter, so you will be feeding hay during that time rather than pasture. Treatment for ketosis is simple and usually involves glucose feedings (you can even do it yourself with molasses if you have the experience).

Tetanus

Although it is more common in horses, sheep can contract tetanus if a rusty metal object injures them. Initial symptoms are a stiff-legged gait and clenching jaws. The bacteria will spread further, causing paralysis that will quickly kill your sheep. Once contracted, you will need to use an antitoxin as well as penicillin to save the animal. You can protect your herd with tetanus vaccinations, which are a good idea because the antibiotic treatment will disqualify your animal from any organic standing.

Bluetongue

The name of this disease is a little misleading because an infected sheep's tongue rarely turns blue, but it will become swollen, along with the lips. Fever and drooling are also evident. You can vaccinate against bluetongue if it is a common threat in your area, but there is no treatment once the animal is ill. It is frequently nonfatal but it can take a sheep months to recover. A small biting fly is usually the culprit in spreading the disease, so insect repellent and bug nets can help keep it from infecting your herd. Insect repellents are externally used products, so there are no ramifications to your organic certification.

Scrapie

The primary symptom of scrapie is that your animals will continually scratch or scrape themselves, which leads to patches of lost fleece. Unfortunately, it is not a simple, treatable skin condition. It is a nervous system disease that causes itching sensations under the skin. After the initial symptoms, your sheep can have seizures and eventually may die. There is no treatment; any animals with symptoms should be killed to prevent the spread of the disease.

Wool

Wool is what makes raising sheep unique, and you will want to know a little about handling a heavy fleece even if you are not specifically keeping sheep for their wool. Sheep kept for meat may need shearing before slaughter, but that will depend on the age you intend to slaughter at. Younger lambs are usually butchered before they need their first shearing. If you keep animals longer than one year, you will want to shear them for their health and comfort in their second year before the hot weather begins.

Shearing

Most shearing is done in the spring, when the weather has warmed up sufficiently to allow your sheep to be comfortable without their heavy coats.

The fleece is not shaved off in a random fashion that leaves you with a pile of woolly pieces. The intent is to create a single piece that naturally holds together in a large pelt. It is best to let a professional shear your sheep for you at least once so that you can see the proper techniques. With experience, an animal can be shorn in about two to three minutes.

The belly of the animal is usually shorn first, and the fleece discarded due to the inevitable dirt and debris from that area. The sides and back will make up the bulk of your overall fleece, but do not leave out the wool from the neck and head; it is all usable.

You want to clip as close to the skin as you can without nicking or harming your sheep.

Typically, electric cutters similar to hair cutters used for humans are used for shearing. A good quality set of electric hand cutters will cost you around $200, and the blades will need to be replaced when they start to get dull. Cutter blades are about $15 each. Depending on how many sheep you have to shear, you can either do your work on a clean floor or save your back with a raised table.

Condition of wool

You will want your wool to be as clean and free of debris as possible in order to get the best price for each fleece. Keeping your pasture free of plants such as burdock will go a long way toward this goal.

Selling raw fleece is the easiest way to handle wool sales, but you will have a more limited market due to the extra effort your customers will end up with. Even clean fleece will have a distinctively oily feel to it from the natural lanolin that is produced by the sheep. This is usually washed off along with any other dirt in the fleece before it is used.

Washing and storage

If you are selling raw fleece, there is little more for you to do with your product. Ideally, you will have customers waiting for their fleece because unwashed wool does not store well due to the oil and dirt. Your wool will have better storage life if you gently wash it before storage. Care must be taken during washing or the fibers will tangle, and you will have useless clump of felt instead of clean wool.

With a washing machine, you can wash a fleece easily, though it will be time-consuming to wash a large number of fleeces. Fill with hot water (not boiling), and add mild detergent or even human shampoo along with your dirty fleece. Let it soak for about half an hour, then run the spin cycle to remove the water. Do not use the regular wash mode. Refill the tub with clean water, soak again, and spin. Your fleece should be much cleaner and will store better for future sales.

Butchering

This will be an inevitable chore for meat sheep and even for milk sheep once your ewes are too old to produce any more milk. Is this something you are going to want to do yourself? Although the size of a sheep or lamb is within reason for a one-man job, trying to manage your own slaughtering facility is tricky. It will have to be inspected by the government if you want to legally sell meat that you have slaughter on your own premises. That can lead to added expenses and paperwork, so most small-scale farmers take their animals to a certified slaughterhouse.

Even if someone else is doing your butchering, you should know about the various cuts of meat so you can tailor your orders to suit your customers' needs and to make sure you are getting the right meat cuts.

Cuts of meat

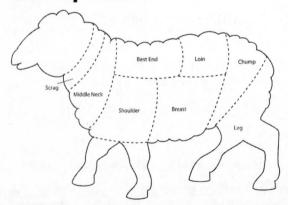

Lamb is the tenderest type of meat you can get from your flock and will likely be the most popular with your customers. Your meat will be considered lamb if you slaughter your young sheep before they are a year old; you can expect to get about 35 pounds of meat from each animal at that age. When they get older, the meat will be stronger tasting and tougher. You will want to slaughter older sheep before they reach 2 years old in order to get meat of a desirable quality. A sheep will give you about half its body weight in meat after it has been slaughtered, though some heavier-bodied breeds will have more meat on them per pound of body weight; you can expect to get about 100 pounds of meat from each sheep.

Mutton and lamb are a common commercially available meat in the United Kingdom and Australia, and the names of the cuts have been adopted from these areas. Traditionally, a sheep is butchered into eight main sections:

- **Scrag:** This is the portion immediately behind the head, at the top of the neck.

- **Middle neck:** This is the neck area and the base of the neck. You get chops and a small roast from the neck.

- **Shoulder:** These are the top portions of the forelegs. Most chops and roasts are cut from the shoulder.

- **Best End:** They are right behind the middle neck, across the back.

- **Loin:** Most of the back region makes up the loin. Rack of lamb is found here, as well as loin chops and the tenderloin.

- **Chump:** This is located behind the loin, just before the hindquarters of the sheep.

- **Leg:** The hindquarters and back legs. Cuts such as steaks and roasts come from the leg area.

- **Breast:** This is roughly the stomach area.

Lamb is tender meat and will not require any additional aging, but if your slaughterhouse has the facilities to let your older mutton age, you should take advantage of it. A two-week aging period before butchering is ideal for tenderizing your meat.

Sheep for Profit

With so many ways to earn profit from your sheep, they can be a lucrative choice for livestock. Whether you are leaning toward milk or meat, there is always a little extra money to be earned with a nice wool fleece as well.

Selling meat

Unless you expect to have many customers for your meat, selling animals by the half or quarter may not make as much sense as it would if you were selling beef. Even though a sheep is smaller, the meat does not have the same demand. Having your animals butchered into individually wrapped cuts is the more typical way of selling meat. One exception would be whole young lambs. Some cultures have traditional lamb roasts for holidays, and that can create some demand for an entire animal at certain times of the year, such as Easter.

When you have your animals slaughtered, the packages are then delivered to you for resale. Keeping them frozen will preserve the meat's quality unless you have customers coming immediately to pick up their meat orders. Then you can just store the immediately purchased cuts in the refrigerator. Lamb or mutton can be frozen for four to six months, though customers may not like the idea of buying six-month-old meat.

Taking this route will require a proper meat seller's license issued by your state's Department of Agriculture. The rules and regulations are not the same across the country, so research the application details. There typically will be a fee for the license, and it will need to be renewed each year.

Selling milk

Sheep's milk is not a common food product, and it is even less known or sought after than goat's milk. Nonetheless, there is a growing market for it, and it can be a viable option as a product on your farm. It is more often used as an ingredient for making gourmet cheese rather than for drinking.

If you are running a milk operation, there is quite a bit you will need to know about in order to have the right facilities for legal milk sales. Your milk needs to be handled properly and stored in the correct containers to keep it fresh and suitable for sale. *More on running a home dairy is covered in Chapter 12.*

To keep your operation legal, you need a license for selling milk. Contact your state's Department of Agriculture for the details and the application process. Depending on your state, you may or may not have to pasteurize your milk before sale in order to keep it legal. Sheep's milk is actually exempt from some of the rules surrounding raw milk in some states, so it would be worth looking into if you were planning on selling your milk unpasteurized.

Selling wool

Multicolored balls of wool

For selling wool, you have a number of options for presenting your product for sale. The easiest, as mentioned in the section on wool, is to sell an unwashed (raw) fleece to customers who are interested in processing the fiber themselves. Just as the organic movement is growing due to a new interest in natural products, more crafters are looking to use raw fibers rather than buying factory-made wool or yarn. Demand for raw wool is growing and should not be overlooked as a sales option. Even so, you will likely find a greater demand for a less primitive version of your fleeces. How much work you want to put into your wool will be up to you.

Once washed, you can card your wool to create a smooth bundle of fibers called a roving. Carding can be done by hand or with a small machine to speed up the process. Rovings are popular with people who prefer to create their own yarn. You can also spin your wool into yarn to create a finished product. Most people who are running a farm and herds of animals prefer to avoid too much additional work to create a saleable product and often stick to selling whole fleece or just rovings.

Depending on your customer base, white wool may sell best because it can be dyed any other color. But some purists enjoy the opportunity to work with wool that is naturally colored in shades of brown, cream, or black rather than dying it.

Although not a food product, yarn or wool can still benefit from an organic label because it lets your buyers know the material is free from chemicals and that your animals were humanely kept.

Sheep make excellent livestock animals that can fit into many different types of farms for different types of income plans, yet they still are seen on small farms less often than pigs or goats. If you are looking for an animal that is a little out of the ordinary, try raising a herd of sheep.

CASE STUDY: MAKING FULL USE OF A SINGLE ANIMAL FARM

Antonio and Molly Manzanares
Shepherd's Lamb
PO Box 307
Tierra Amarilla, NM 87575
www.organiclamb.com

The Manzanares raise only sheep at Shepherd's Lamb, but they have diversified their product offerings to maximize their earnings from the flock. They sell meat, sheepskin, raw wool fleece, carded wool, and even hand-dyed yarn. Sales are made through the local farmers market, and they sell a portion of their harvest to local natural food stores and restaurants. Lambs are sold either whole or half to customers during the fall season.

Their farm is larger than most with nearly 1,000 ewes, but they started out with only 40 back in the early 1980s. They achieved organic certification in 1998, and owner Molly Manzanares says the process went fine. But she also says that changes to the system have added extra hoops to jump through since then.

The sheep graze on pasture, but they feed with purchased organic hay for about three months of the year. Sometimes, it can be difficult to get organic hay at a decent price during those months.

Their farm manages a small staff with a farm hand and shepherd, and they are looking to expand to hire a marketing employee as well. All butchering is done outside the farm at a slaughterhouse in Colorado, so they do have to handle a lot of transportation with their animals.

Over the years of growth that the farm has seen, Molly says there is nothing she would do differently if she had the chance. To new farmers, she warns that there is a lot of hard work involved, and you are better off if you have a good partner to work with.

Chapter 11

Other Livestock Animals

So far, this book has covered the most common animals used in small-scale livestock farming, but they are by no means your only choices. For anyone wanting to raise something different, either just for the interest in it or to help set yourself apart from your competition, some of these other animals may be what you are looking for.

This information is just an overview of each animal type, and you will want to do further research to get an accurate picture of the complete care requirements for any of them.

Rabbits

Rabbits are the ultimate in small-scale farming and will fit in with even the smallest of operations. They will produce a great deal of meat in a small area, and their manure will create excellent fertilizer for your garden. Each animal may only produce a few pounds of meat, but their amazing reproductive capabilities will make up for that in sheer volume.

Many people who raise rabbits for meat will keep their animals exclusively in wire cages, which is not ideal for anyone with an organic farm in mind. The treatment of the rabbits is as important as the food you feed them, so plan on an enclosure system includes outside areas as well as a safe indoor area. Place a nesting box in your female rabbits' living space, and arrange to have cages or hutches separate for the males. As with most other animals, you do not want your males and females together all of the time. The real danger is actually from the females. Once pregnant, a female rabbit (doe) can be aggressive about her territory and can kill your males.

A doe will produce a litter of six to ten baby rabbits in about one month. They can be weaned at two months, at which time you can rebreed the female. At this pace, you can easily produce a large number of animals in a short period. Continue to feed and raise the young for another two to three months. They should be around 5 pounds by the time they are ready to be slaughtered.

Rabbit meat is lean and mild tasting. It is often compared to chicken. Given the small size of the animals, the slaughtering chores are much easier to manage compared to butchering a sheep or a cow. The downside to this is that most certified slaughterhouses do not handle rabbits, which leaves the task completely on your shoulders. This is where a legal loophole works in your favor. Rabbits are not regulated like other livestock animals, and you are not required to have a government-inspected facility to sell rabbit legally. Check with your state's Department of Agriculture for details.

For feed, you will need to supply your rabbits with good quality hay (alfalfa or Timothy) as well as a constant supply of fresh water. Water bottles with a ball valve at the bottom (such as those you would see in a pet store) work well for rabbits and the water stays much cleaner. Along with the hay, you can add extra nutrition with fruits, vegetables, and other greens, such as dandelion.

Alpacas

An alpaca is a grazing animal, similar to a llama but smaller. They are raised exclusively for their fiber, which is spun into soft wool. When compared to sheep's wool, the fleece from an alpaca is softer, lighter, and is not coated with lanolin. It is much easier to clean and use, so there are definitely some aspects of keeping alpacas that sets them apart from sheep. Their temperament is gentle, and many farmers prefer to keep them when there are children around the farm.

The raising of alpacas is similar to raising other animals like sheep or goats. You can easily house them in a typical barn, and about eight to ten alpacas can share an acre of pasture space. Feeding an alpaca is no different than feeding a sheep. Provide decent pasture, additional hay, and some grain.

You can start breeding your female alpacas when they are 18 months old, and the gestation period is just under one year. Even after such a long pregnancy, alpacas will typically produce just one young (called a cria) at a time. Although the return may be low, alpacas can be bred any time, and you do not have to keep track of heat cycles or seasonal changes.

Your profit will be limited to fleece, as alpaca meat is not customarily eaten; however, the demand for unique fibers is growing and there is success to be had with these animals.

Emus and Ostriches

These two huge birds may not seem like potential livestock animals, but demand for their meat, eggs, and feathers is growing enough to create a market for them. Their fat can even be rendered into saleable oil. They are similar species of birds, though the emu is smaller than the ostrich. In terms of livestock, they are usually classified under poultry and can be subject to the same regulations as chickens.

Even if you are relatively familiar with most other four-legged animals, these large birds will be a change of pace. Emus tend to

be more common in the livestock trade because they are a more manageable size compared to ostriches. An adult emu will be 6 feet tall and can weigh about 150 pounds. Although they are birds, they are completely flightless, so you do not have to keep them in a covered enclosure.

A standard fenced pasture will suit fine, though it will need to be at least 6 feet high.

Emus will be around 2 to 3 years old before they start to lay eggs, and they will produce roughly two eggs per week. Their eggs will be about the size of a grapefruit, and the shells are dark green. Fertile eggs will hatch in about 50 days. When breeding your emus, do not forget that their large eggs will not fit in a standard egg incubator.

They are generally docile animals, but their long, muscled legs can deliver quite a kick if they decide to get aggressive. Emus are not suitable for a complete animal novice. The regulations surrounding their use and sale will vary by state. Some states have these birds — usually referred to by their family name of "ratites" — listed under poultry for licensing and inspection regulations, but some consider them exotic game animals that will require more paperwork.

Emu and ostrich meat is much more like red meat in texture and taste than poultry. Slaughtering tasks will likely be your own because most commercial operations will not know how to handle

birds like this. A good book on ratites should guide you through the process well enough.

You can keep emus on pasture, but there is commercial emu feed on the market. Finding organic sources may be a challenge. With a little knowledge, you can likely avoid the need for premade foods.

Because these birds are different from other animals, research their needs thoroughly before buying any chicks.

Ducks and Other Poultry

Chickens were discussed in length in Chapter 6, but they are not the only poultry that you could consider raising on your farm. Ducks, turkeys, geese, guineas, and other game birds are all other possibilities. The basics of their care are all similar to chickens, so if you are familiar with them, you are well on your way to branching out into other birds. Just keep in mind that they all have their own distinct needs and requirements.

All poultry can be raised for either eggs or meat (or both). Ducks, turkeys, and geese are larger than chickens and typically produce larger eggs and more meat per bird. They will need more space for pasture and housing — and more food. Ducks need about twice the area as a chicken, and geese need more than double the room that a duck needs.

If you are going to raise ducks, you have to provide them with some area with water they can swim in. They need at least a small amount to paddle around in so they can properly bathe and keep their feathers clean. A kiddie pool will do for a few ducks, but a pond works best because they can be quite dirty in the water and a small volume will need to be changed frequently. Ducks will lay their eggs at a frequency similar to chickens, about three to six per week. The eggs will be a little larger but should still work in an incubator if you are raising fertile eggs.

Turkeys will give you eggs roughly every other day, but they are more commonly raised for meat than for eggs. Because they will grow considerably larger than chickens, they are not slaughtered quite so early. Expect to have your turkeys around for at least four months before butchering time.

Guineas are more flighty than other birds and may need a covered pen to keep them from flying out of their enclosure. They are known for their loud voices. They are less well known than the others, but they can be raised just like chickens.

With nearly any kind of poultry, the meat and eggs are familiar enough that you should have no problem finding customers just as though you were dealing with chickens.

Buffalo

Even the realm of cattle has a few alternatives if you want to raise something other than cows. Buffalo, also called bison, are starting to make an impact on the beef industry. Hybrids between

cattle and buffalo are another option, called "beefalo." Water buffalo are also seen in meat-producing operations and should not be confused with the bison. Water buffalo and bison are two different animals.

Buffalo will grow larger than a cow. They are not truly domesticated animals, so take care when walking around in your pastures. That is not to say they are aggressive or dangerous — they are just not as docile as a typical cow would be. The positive side is that they are hardier and can survive in harsher conditions with less intervention from you as the farmer. They thrive on pasture and do not usually need additional supplements or grain.

The meat from a buffalo is much like regular beef and should sell well to customers looking for red meat. Because even the females will easily top 1,000 pounds in size, each animal will produce a lot of meat. Large slaughterhouses may be able to take care of your butchering needs, but smaller ones may not. The animals are similar enough to cows that a facility used to slaughter cattle should have little difficulty with them.

Compared to cows, buffalo grow more slowly. You will not be able to start breeding your females until they are about two years old. Once impregnated, she will have her single calf about nine and a half months later. They grow to their full mature size after five years, so slaughter is usually done at 3 to 4 years old.

Bees

Although it may not be truly accurate to consider bees as livestock, they are certainly capable of turning a profit on your farm and should be included.

They are a bit of an odd option for an organic farm, and there are a few loopholes in the organic standards that bees and honey fall into. The catch with raising bees is that you cannot control or monitor where your bees are actually feeding. They can (and will) fly up to three miles to gather nectar and pollen. It may seem like this would make them ineligible for organic certification but that is not the case. Honey falls outside of the regulations. Anyone can label his or her honey as organic.

Most people raise bees for their honey but you can add to your income stream by selling beeswax as well. How much honey you can harvest each year will depend on the size of your hives and how many bees you actually have in your population. A busy hive can give you up to 100 pounds of honey each year, on average.

Keeping bees is nothing like keeping other animals. There is not likely to be too much overlap in terms of equipment or supplies that you will already have on hand if you want to get started. You will need a hive structure, coveralls, a hat, a netted veil, and a bee smoker. You will also need a source to buy your first swarm of bees, and most suppliers will deliver.

A healthy colony of bees will be made up mostly of worker bees and one queen bee. If something happens to your queen, you must get a new one or the colony will fall apart and possibly swarm out of the hive. Only the queen lays eggs, and the workers make the honey to feed the offspring when they hatch.

Within the hive structure, there are frames where the bees lay out the beeswax to store their honey. Once a year, usually during late summer, you can remove these frames and extract the honey. A hot knife is used to slice the top layer of wax from the honeycomb, and the honey is poured out. There is actually little work involved for you. Your bees do not require any feeding throughout the year either.

Understanding how a colony of bees operates is important when raising bees for honey, so more thorough research is recommended before beginning.

Chapter 12

The Home Dairy

iven the complexities of managing a dairy operation, a chapter is being devoted strictly to this one aspect of your farm. Whether you are raising cattle, goats, or sheep, this is what you need to know for your milking plans.

Goat, Sheep, or Cow Milk?

They may seem similar, but there are some significant differences between the various dairy animals you could raise. Understanding your product will go a long way to understanding your future customers.

One big issue is size and quantity. A cow will provide you with a great deal more milk than either a sheep or a goat, but you will have to compensate for that extra milk by managing a larger animal with larger feed and space requirements.

Aside from the quantity concerns, the biggest difference between these three milk types is customer acceptance. In North America, cow milk is by far the most commonly used milk, and it generally will sell more quickly than the others.

The milk itself is different from one animal to the other as well. Goat milk has proteins in it that are more similar to those in human milk when compared to cow milk, which means some people with milk allergies are able to drink goat milk but not cow milk. This is also true for most people who have lactose intolerance. The flavor will be different from one kind of milk to another. Some people feel they all taste more or less the same, but others will notice a stronger flavor to goat or sheep milk.

Facilities

Whether you need a dedicated building for milking will depend on the size of your herd and how much space you have to spare. If you cannot afford the space or the cost, you can use an area in your barn instead. A clean stall or extra room will usually be enough to accommodate your milking equipment, and you should have room for one or two animals at a time.

It is not a good idea to try to milk your animals in a communal pen, even if you are hand milking. Your dairy animal needs to be relaxed in order to release her milk, and that can be harder to achieve with other livestock milling around. It is also hard to keep the area clean.

Your milking area should also include a feeder so you can give your animals something to eat during milking. Keeping them occupied while you milk will make a big difference in their cooperation. Cows are usually fine to stand still with a trough, but you may want to use a milking stand for smaller animals such as goats or sheep.

Small-scale milking equipment

Here is where you are going to have the most options and choices when you set up your dairy operation. You can keep it simple

and milk by hand into a bucket, or you can have high-tech milking machines. Deciding which way to go will depend on many factors: your budget, how much space you have, how much time you have to spend on milking, and how many animals you have to milk each day.

To know what equipment you will need, you need to decide how you want to milk. In terms of time, it will take about 20 minutes to hand milk a cow (somewhat less for goats or sheep) or five minutes with a machine. With more than two or three cows, this time difference can really add up.

If you choose to stick with hand milking, the only equipment you will need is a selection of stainless steel buckets and a stool. Machine milkers obviously will be more complicated.

When it comes to machines, you can purchase single-animal units that would suit a small-scale farm. Costs will depend on the features and brand, but expect to pay between $1,000 and $2,000 for a single-animal unit. There are different models for cows, sheep, and goats that still run in the same price range. Larger machines can handle two, four, or more animals simultaneously.

Milking machines will vary by model, but they all have the same general function. A hose with a teat cup at the end is attached over each teat on the udder (four for cows, two for goats or sheep), and suction is created when the machine is turned on. It creates a pulsing suction similar to how you would milk by hand. The expressed milk is collected in a central canister or bucket.

Once collected, there are equipment requirements for storage. Any large canisters will do as long as they can be thoroughly cleaned and sanitized between uses. Stainless steel is the most common material, and large cans made for milk can be purchased.

Aside from the actual milking equipment itself, you will need a few other items and supplies to run an efficient home dairy. You will need to wipe off and clean your animals' udders and teats with each milking. A simple bucket with warm soapy water and a rag or paper towel will do fine; you can even go one step further and use a teat disinfectant. *There will be more information on proper udder care later in this chapter.*

Another helpful tool to have around the milking area is a pair of hobbles or kickers. Their exact name can vary by region. They are simple devices that you put between your animal's back feet to keep them from accidentally (or intentionally) knocking over your milk bucket.

Once the milking is completed, the next steps will require additional equipment. Filter your milk before storing it to make sure there are no pieces of debris or hair in it. Disposable milk filters can be purchased from most dairy supply companies. They are used with stainless steel strainers to keep your milk clean.

You may also need a pasteurizer, though this will not be necessary if you are selling raw milk. *The legal issues surrounding the sale of raw milk were discussed in Chapter 2.* A pasteurizer will heat the milk to at least 145 degrees Fahrenheit to kill any bacteria. Depending on the type of machine, it may heat the milk hotter but for a shorter period. Small units that can pasteurize 2 gallons at a time are available on the market. Once heated, the milk is quickly cooled back down and then refrigerated for sale.

Milking stands

Milking stands are used for smaller animals to raise them up off the ground so their udders are at a more comfortable height for the milker. It may not matter much if you are using machines, but it will make a difference if you hand milk.

They are simple constructions similar to a low table. Steps or a ramp can allow the goat to walk up on their own, which makes it easier to use than trying to lift your animals each time. Two vertical boards set apart just enough to allow the head through at the front will restrain your goat's or sheep's head. This is called a stanchion. With a feed bucket on the other side of the stanchion, your goat or sheep will usually stay still while you milk.

Cleanliness and udder care

Being handled, by machines or by hand, twice a day can take its toll on your animals' teats and udders. Proper hygiene and care will help to prevent mastitis as well as keep your cow, goat, or sheep happy and comfortable. Chapped teats, on the other hand, will irritate your animal and make milking difficult.

Keeping the udders clean is the most important part of milking, not only to keep the milk itself clean but to also prevent the spread of disease. With every milking, you will want to wash the udders and teats with warm water. If you want to be extra careful, disinfectants formulated specifically for teats also are available on the market. Some contain iodine while others contain peroxide. Neither will have any impact on your organic standing; so choose at your discretion. Iodine can be a little harsh and may result in dry or chapped teats. Milder organic formulas can be found if you prefer something with fewer chemicals. You can either wipe the teats down or actually dip them in a bottle of the disinfectant. For added convenience, there are premoistened wipes that work like baby wipes as well.

If the skin on the teats starts to get dry, it can get cracked or chapped, which will make your animal uncomfortable and lead to problematic milking sessions. If it gets bad enough, you are risking infections or further injury. Even without disease, a teat that bleeds during milking is not going to be good for your business. Many different kinds of udder cream or "bag balm" can be used to keep the skin moisturized and conditioned.

Because mastitis is always a possible threat with any dairy animal, you will want to test for it periodically so that you do not risk selling any infected milk. There are test kits you can purchase, and milking into a small strainer can show if there are any clots in the milk. By checking regularly, you can catch a case of mastitis quickly and treat it immediately.

Milk care and storage

The most important thing you need to worry about with fresh milk is cleanliness. Keep all your buckets and jars clean, and clean your animals' udders before every milking. Immediately after collecting the milk, it should be strained or filtered to remove any dust, hair, or other debris that has gotten into the container. This is seldom a problem with milking machines because the collecting canisters are not open to the air. After straining, you will want to either refrigerate it immediately or process it with a pasteurizer first.

Regulations and organic guidelines

Raw milk may or may not be legal to sell in your state, and that is generally the main legal issue with trying to sell milk from your own farm. *Review the Legalities section of Chapter 2.*

When selling milk, you will need the proper license, as regulated by your state's Department of Agriculture — the specific rules, regulations, and hygiene standards will be specific to your state. Licenses are issued annually and, typically, have a fee associated with them along with inspection requirements.

As for the organic side of your dairy, the regulations are pretty much the same as for selling meat. You can only feed your dairy animals certified organic food, and you cannot use any growth hormones or antibiotics when raising your animals. *A more complete description of organic standards can be found in Chapter 2.*

Chapter 13

Marketing

Once you have your operation in place and have a product to sell, your next big job will be to find those customers. Starting a new farm business can be daunting, and how you market yourself is a big part of your future success.

Who are Your Customers?

You will not know where to look for customers until you firmly establish who you are looking for. You likely considered this when you first thought about your farm and what animals you wanted to raise. Now you need to give it further thought to build a plan to reach those customers.

Regardless of what animals you are raising and the specific types of products you are selling, your main customer base will be people who are concerned about the quality of their food — people who want healthy, natural foods without the added chemicals found in mainstream farming practices. These people do not mind spending a little more than standard supermarket prices for the organic products they are looking for.

Finding your niche

If you are lucky, you will find yourself with a large local customer base with little competition. Some regions do not have a lot of options for organic meats, so this is a possibility.

Be prepared to find a smaller niche so you can target your advertising to a select group of people in order to gain the advantage over any similar businesses in your area. Finding this niche can present a challenge, so allow yourself some time to work on it. In a sense, you already are working within the niche of organic food. But if you are competing for the same customers with others selling similar organic products, you need a more specialized approach.

For example, you could market your meat specifically for the barbecue and target the grill enthusiast, or use your advertising to let people know your goat milk is fine for lactose intolerant people.

Where to Sell Your Products

Small-scale farmers typically sell their produce locally, and you can take advantage of several venues to maximize your sales. Some opportunities will be obvious, but do not rule out any that are a little out of the ordinary. You never know where you biggest sales will come from.

Farmers markets

This is likely one venue small farmers think of first when starting to plan how to sell their livestock. Small towns and big cities have them — sometimes even several. These types of markets can be the lifeblood for a small-scale farmer and will give you an opportunity to present your products to a range of potential customers who are already on the lookout for locally grown/raised food.

These types of markets are sometimes informal, with people selling items from small tables in a parking lot, or they can be more  structured events where you have to pay a fee to rent a space or a vendor's booth. Markets where you can sell free are the safest to try because you will have no overhead to cover. Any market that charges you should be investigated a few times before you commit to paying for a space. Visit during the day as a customer to see how busy the area is, and watch closely to see how many people are browsing and who is actually buying. Then you can decide if it is worthwhile to pay for a space. This research may not be that important if the fee is small, but large markets may only sell vending spaces for the entire season, and you will want to judge the sales potential before making that kind of commitment.

As a seller of meat, milk, or eggs, you will be at a disadvantage compared to the others who will be selling fruits or vegetables due to the perishable nature of your products. An indoor market may be able to supply electrical power you can use to run a refrigerator or a freezer, and even some outdoor markets may have the same feature. Using a small gas generator is another option, though it can be a noisy distraction if you do not have the space to set it apart from your stall.

Even if you feel you cannot comfortably sell your perishable meat or milk on a hot day, you can still use a farmers market stall to promote your farm. Posters and information about your animals and products can help draw customers to your farm where they can purchase cuts of meat or even sign up to buy half an animal. Always have fliers or business cards available so your customers can find you again.

Farm sales

You do not necessarily have to go off site to sell your products. Selling directly from your farm is a common method, though it will mean more advertising is necessary to let people know who you are and what you are selling. This

This farmer sells his extra hens to the community from his farm.

can be convenient as you can set your own hours, and there are no transportation or shipping costs to contend with.

When selling any product from your farm, you will want to present a reasonably clean and professional-looking area for your customers. Having a room set aside for sales is excellent if you can afford the space, though it is not necessary to set up a storefront in order to sell this way. Many customers like the chance to not only buy organic produce, but also to see where the animals are raised.

CSAs

CSAs (community supported agriculture) are becoming a popular way for small farmers to sell to their customers. The idea is that people purchase a share in a farmer's harvest for the year, which is delivered weekly throughout the growing season, so the buyer gets fresh, in-season produce all season long as plants are harvested. This may not be suitable for a meat operation that has one main slaughtering period in the fall, but the idea is popular, and you can always get creative in making it work for your farm. If you are selling milk and/or eggs, this may be a good option because you can supply these products on a weekly or biweekly basis for delivery.

Another way to make use of the CSA principle would be to work with other farmers who sell shares of fruit or vegetable produce. They can provide their harvest through the year, and you can add your additional meat items when it is the right season. Such a partnership would add extra value to the other farmers' CSA offerings and would free you from needing to have a product all year long.

Auctions

Your meat, milk, or egg products are not sold in this way, but you can use livestock auctions, as long as there is such an auction within a reasonable distance, to help sell your live animals. These can be animals you are getting rid of

Sheep at an auction house

because they are not needed (excess males in a dairy operation, for example) or those that have been raised for meat but you prefer to sell "on the hoof" rather than slaughter yourself. You will need a truck and livestock trailer to move your animals to the stockyard. Selling live animals this way can be preferable to selling cut meat because each sale is larger. On the other hand, your pool of potential customers will be substantially smaller because the average consumer does not buy animals for meat this way.

An exception is the sale of horses. Because you are not selling them for meat in the first place, selling horses live at an auction is a common way to make your sales.

Selling to stores

You can get larger sales if you take a different angle and sell your products to local stores for resale to customers. This wholesaling approach can be appealing, particularly if you have nearby stores that are interested in locally raised food products. Although this is definitely something to consider, you will find that your licensing requirements will change if you are want to develop products for resale. Selling directly to the customer is a simpler process with less paperwork.

Mail order

Believe it or not, you can also sell your products globally if you want to expand your customer base. This is not going to be a suitable venue if you are selling eggs because of the difficulty in securely shipping eggs. Milk can be managed if you have tightly sealed containers. Meat would work the best in terms of surviving the rigors of shipping.

In any case, you have to keep your products cold. Overnight shipping along with some dry ice is a common method for selling frozen foods and can be managed even by the small farmer.

This route would mean you can advertise anywhere and even take full advantage of the Internet to sell your products anywhere in the world. Proper shipping containers can add to your overhead, but having access to customers outside your own region can make up for this added expense. You can ship your products across state lines as long as they are USDA approved, such as at a certified slaughterhouse.

Promoting Your Business

This can be a difficult part of a business if you are foremost a farmer, not a businessperson. But without proper promotion, you will not be able to reach potential customers. Word-of-mouth is a powerful tool, but do not rely on it alone to let people know about your farm.

Maintaining a website

Even if you are only selling your produce at the local farmers market, and you have no interest in trying to drum up business anywhere else, a website is still a valuable tool for your farm. More people are turning to the Internet to find things, even when they are looking for something in their own town. You want *any* potential customer to find you.

In this day and age, people who find you via your website will not necessarily assume or expect you to offer mail-order sales through your site. Your website can provide an attractive and informative look at your farm, describe your organic principles, and let people know where and when they can purchase your products. Consider it an online brochure to promote what you are doing.

Having a website is not as complicated or expensive as you may think, though you should only tackle the job yourself if you have a reasonable understanding of computers. With a book or two on Web page design, you could probably do it yourself. An introductory book, such as *Creating Web Pages for Dummies,* would be a good place to start.

If you do have a professional create a site for you, do not let him or her convince you that it will take hundreds of dollars to create or maintain. This should be a relatively inexpensive part of your advertising plan.

When using a website with your business, provide both e-mail as well as a phone number so people can contact you. Even if you are not the kind of person to use the Internet or computers yourself, make the effort to check your e-mail regularly. Potential customers who are using the Internet will expect a quick response simply because instant communication is the nature of the system. Checking e-mail once a week is not often enough; daily is best.

Advertising

Of course, you will also want to promote your business through more conventional advertising channels as well. Most people create a name for their farm, and that becomes the brand name you want to promote. Use your name on business cards, flyers, brochures, and any signs you use for your farm sales or farmers market table. You are primarily looking for local customers, so you will want to focus your promotion efforts on your own community.

Local advertising

You can use any of the standard forms of advertising that would suit a small business, such as an ad in your local newspaper, flyers posted at any community bulletin board, and a noticeable sign at your farm's location.

Taking advantage of local events is another way of getting your name out to prospective customers. Offer to supply food either free or at a reduced cost for an event in exchange for a small sign promoting your contribution. Not only does this approach give you a place to display your name, it also generates goodwill in the community and even gives people the opportunity to actually taste your products.

Hosting an open house at your farm can also generate some interest in your operations and let customers see everything that you do. Of course, you will want to have some samples available for all the patrons who attend.

Online and social networking

Having a website has already been mentioned as a way of having an online presence, but it is not the only way to advertise on the Internet. The advent of social networking has created a new arena where you can promote a small business directly to an interested audience. Social networking can be confusing to anyone not familiar with the Internet and how it has developed over the past few years. You do not have to master dozens of sites in order to take advantage of this new media. A Facebook™ account and a Twitter™ account are sufficient to start promoting your business. Many people who read your posts will be too far away to actually become customers, but do not let that deter you from using this option for advertising. Even if just a handful of people in your own area happen across your postings, you can benefit.

Full details of these websites are too elaborate to detail in this book, but there are many good online tutorials and even social networking books that can help you get started. Both Facebook and Twitter allow you to build an audience of followers who will then see your latest posts or comments. By following others, you can slowly build up an interest in your own business. Offer helpful information on organic foods, livestock, and even healthy cooking to get people hooked on what you have to say. Add in a few mentions now and again about your farm and your products, and it becomes a great way to promote yourself without resorting to 100-percent advertising.

Conclusion

Raising livestock for profit can be difficult and challenging, and there is no shortage of physical labor involved. Add in the necessary steps needed to have an organic farm, and you have quite a job ahead of you. Even so, there is nothing like the satisfaction of producing healthy and wholesome food for your family and your customers.

Not only are you now more familiar with the care needed for various livestock animals and all the requirements of organic certification, but also the case studies have given you a personal glimpse of what life on a moneymaking organic farm is like.

Once you understand the nature of organic standards and certification, many farmers find it no more difficult to manage an organic farm than a conventional one. Do not let the concept of going organic deter you from running your farm in the way you want. Organic practices are not only better for your animals and your customers, but they also are a better way of farming for the environment as well.

Glossary

barrow: A castrated male pig; also sometimes called a stag if the castration takes place after puberty

bleat: The call or sound that a goat makes

boar: A male pig that has matured to breeding age

broodiness: The instinct that a hen has to sit on her eggs once laid

broodmare: A female horse whose sole purpose is for breeding, rather than racing or working

buck: A male goat; also called a billy

bull: A male cow of any age

calf: A general term for any young cow

chevon: The proper term for goat meat

chick: A general term for young chickens of either sex

colt: A younger male horse

combs: The fleshy growths on top of a rooster's head

cow: A female cow that is older than 2 years or that has had at least one calf

doe: A female goat; also called a nanny

ewe: A female sheep of breeding age

farrier: A professional who shoes horses and cares for hooves

farrow: To give birth (for a pig)

filly: A younger female horse

foal: A young horse of either gender that is still with its mother

foaling: To give birth (for a horse)

freshening: The act of impregnating an animal to replenish her supply of milk

gelding: A male horse that has been castrated

gilt: A female pig young enough to have not yet been bred

heifer: A young female cow under 2 years old

hen: A female chicken that has begun to lay eggs

kid: A young goat

kidding: To give birth (for a goat)

lambs: Young sheep

lambing: To give birth (for a sheep)

mare: Female horse of breeding age

mutton: A term used for goat meat, though some regions

use this term for meat from sheep

polled: Naturally having no horns; used for any animal

pullet: A young female chicken, before she begins to lay (usually under a year)

ram: A male sheep of breeding age

rooster: A male chicken of breeding age; also sometimes called a cockerel

shoat: A young pig, usually still unweaned; also called a piglet

sow: A female pig that has had her first litter

stallion: A male horse of breeding age

steer: A male cow that has been castrated, typical in beef production

vent: Single body opening for a chicken to excrete waste

wether: A castrated male goat or sheep

Bibliography

Aubrey, Sarah Beth. *Starting & Running Your Own Small Farm Business*. North Adams, Massachusetts: Storey Publishing, 2008.

Bennett, Bob. *Storey's Guide to Raising Rabbits*. North Adams, Massachusetts: Storey Publishing, 2009.

"Common Chicken Diseases." *Avianweb.com*. **www.avianweb.com/chickendiseases.html**.

Damerow, Gail, ed. *Barnyard in Your Backyard: A Beginner's Guide to Raising Chickens, Ducks, Geese, Rabbits, Goats, Sheep and Cattle*. North Adams, Massachusetts: Storey Publishing, 2002.

Ekarius, Carol. *Small-Scale Livestock Farming: A Grass-Based Approach for Health, Sustainability and Profit*. North Adams, Massachusetts: Storey Publishing, 1999.

Emery, Carla. *Encyclopedia of Country Living*. 9th ed. Seattle, Washington: Sasquatch Books, 1998.

Family Milk Cow Dairy Supply Store. *Familymilkcow.com*. **www.familymilkcow.com**.

"Horse Breeds: Basic Descriptions of Widespread Equine Breeds." *Horses-and-horse-information.com*. **www.horses-and-horse-information.com/horsebreeds.shtml**.

"Industry Statistics and Projected Growth." Organic Trade Association, *Ota.com*. **www.ota.com/organic/mt/business.html**.

Nelson, Melissa. *Complete Guide to Small-Scale Farming*. Ocala, Florida: Atlantic Publishing Group, 2010.

"Shearing." *Sheep101.info*. **http://www.sheep101. info/201/shearing.html**.

Storey, John and Martha. *Storey's Basic Country Skills*. North Adams, Massachusetts: Storey Publishing, 1999.

Thomas, Steven and George Looby, DVM. *Backyard Livestock: Raising Good, Natural Food for Your Family*. 3rd ed. Woodstock, Vermont: Countryman Press, 2007.

U.S. Small Business Administration. *SBA.gov*. **www.sba.gov**.

USDA Food Safety and Inspection Service. *Fsis.usda.gov*. **www.fsis.usda.gov**.

Weaver, Sue. "Meet the Money Makers." *Hobbyfarms.com*. **www.hobbyfarms.com/livestock-and-pets/raising-meat-goats-25028.aspx**.

"Weighing a Pig Without a Scale." *ThePigSite.com*. **www.thepigsite.com/articles/541/weighing-a-pig-without-a-scale**.

Author Biography

Terri is living on five rural acres and slowly building it into a thriving organic farm. Although she grew up in the city, today she prefers a natural country life with her significant other and young daughter. By managing a freelance writing career from home, she can spend most of her time outdoors tending to both garden and live-stock, which has given her a great deal of hands-on experience that she treasures.

Not only has she gleaned her knowledge from experience, but also she has learned from the numerous helpful neighbors who seem to constantly have tips and suggestions to offer. Eventually, Terri hopes to be more self-sufficient and produce most of her own food. When not in the garden or barnyard, she studies gene-alogy and collects antique typewriters.

Index